P9-CKV-289

HOME
ELECTRICAL
REPAIR AND
MAINTENANCE

HOME
ELECTRICAL REPAIR AND MAINTENANCE

James L. Kittle

McGraw-Hill Book Company

New York St. Louis San Francisco Auckland Bogotá Guatemala Hamburg Johannesburg Lisbon London Madrid Mexico Montreal New Delhi Panama Paris San Juan São Paulo Singapore Sydney Tokyo Toronto

HOME ELECTRICAL REPAIR AND MAINTENANCE

McGraw-Hill/VTX Series.

1 2 3 4 5 6 7 8 9 10 D O C D O C 8 9 3 2 1 0 9 8 7 6 5

ISBN 0-07-034899-5

LIBRARY OF CONGRESS CATALOGING IN PUBLICATION DATA

Kittle, James L., (*date*)
Home electrical repair and maintenance.
Includes index.
1. Electric wiring, Interior—Amateurs' manuals.
I. Title.
TK9901.K53 1985 621.319'24'0288 84-23347
ISBN 0-07-034899-5

Sponsoring editor: Jeffrey McCartney
Editing supervisor: Margery Luhrs
Designer: M.R.P. Design

Trademarks

Below is a list of registered trademarks used in this book, along with their corresponding companies.

ACE	Henry L. Hanson, Inc.
AMP	Amp, Inc.
AMPROBE	Amprobe Instrument, Division of Core Industries, Inc.
BLACK & DECKER	Black & Decker (US), Inc.
BRIGHT STAR	Bright Star Industries, Subsidiary of Kidde, Inc.
CHANNELLOCK	Channellock, Inc.
CRAFTSMAN	Sears Roebuck & Company
CRESCENT	Crescent Tool Division of Cooper Tools
DIAMALLOY	Diamond Tool and Horseshoe Company
EICO	Eico Electronic Instruments, Inc.
FUSTAT FUSTRON TRON	Bussmann Manufacturing, Division of McGraw-Edison Company
GRAINGER (DAYTON)	Grainger, Inc.

LEVITON	Leviton Manufacturing Company, Inc.
GREENLEE	Greenlee Tool Company, Division of Ex-cell-o Company
HANSON	Henry L. Hanson, Inc.
HONEYWELL	Honeywell, Inc.
IDEAL	Ideal Industries, Inc.
ITT HOLUB	ITT Holub Industries, Division of International Telephone and Telegraph Corporation
KLEIN TOOLS	Klein Tools, Inc.
MOLLY ANCHORS	Molly, Division of Emhart Fastener Group
MONTGOMERY WARD	Montgomery Ward
NATIONAL ELECTRICAL CODE (NEC)	National Fire Protection Association, Inc.
QUICK-WEDGE	Kedman Manufacturing Company
RIDGID	The Ridge Tool Company, Subsidiary of Emerson Electric Company
ROMEX	Rome Cable Company
RAWLPLUG	The Rawlplug Company, Inc.
SEALTITE	The Anaconda Company, Brass Division
SKIL	The Skil Corporation, Subsidiary of Emerson Electric Company
SQUARE D	Square D Company
TOASTMASTER	Toastmaster Products, Division of McGraw-Edison Company
TRUE-VALUE	True-Value Hardware Stores
TY-RAP	T & B Manufacturing Company
UL LABEL	Underwriters Laboratory, Inc.
VISE-GRIP	Petersen Manufacturing Company
G. E. WAFFLE IRON	General Electric Company, Appliance Division

To my special friend, Clara Antoinette

Contents

CHAPTER 3 Hand Tools for Wiring 35

CHAPTER 4 Wiring Materials and Devices 61

Acknowledgments

To mention every person who has, in some way, made this book a reality is impossible. The two journeymen who taught me the fundamentals of the trade, the Schroeder brothers, who owned the business, were the best teachers a young apprentice could have. They taught me how to do it right and to have pride in my workmanship.

The following companies have kindly provided text and illustrations for this book. Their cooperation and courtesy are greatly appreciated: Enerpac, Division of Applied Power, Butler, Wisconsin; Klein Tools, Inc., Chicago, Illinois; Bryant Wiring Devices, Division of Westinghouse, Bridgeport, Connecticut; AFC/A Nortek Company, New Bedford, Massachusetts; and City of Troy, Troy, Michigan, Jay N. Winslow, Chief Electrical Inspector.

I would also like to thank my editor at McGraw-Hill, Jeffrey McCartney, for his help and good advice in the preparation of the manuscript; Ralph Brande, with whom I discussed the arrangement of the chapters; Margery Luhrs, my editing supervisor, who worked with me on finalizing the manuscript; and all the staff of McGraw-Hill.

And last but most important, I thank my special friend, Clara Antoinette, who put up with me during the writing and rewriting of the manuscript.

Preface

There are many how-to books in every field of endeavor. This is especially true in the mechanical trades such as plumbing and heating, carpentry, and electrical wiring. As a licensed journeyman electrician, I have used my many years of experience in all phases of electrical wiring to make this book on the repair and maintenance of electrical wiring practical and informative.

I have personally used and worked with all procedures and methods described in this book. Some procedures I have developed to make the work easier, faster, and more efficient. The National Electrical Code® (NEC) is followed, and all procedures are in accordance with it.

Many people install wiring in their homes and other buildings without the knowledge of or regard for correct procedures and safety precautions. As a result, sometimes they run into trouble and start a fire. Electrical wiring is an exacting procedure having certain rules and ordinances that must be followed. The National Electrical Code is the electrician's Bible. It provides for the safety of all electricians and their clients by outlining a set of guidelines for standard procedures covering every possible wiring situation.

This book can guide you through all phases of repair and maintenance with complete explanations of the methods used. It is my aim to train you, the beginning electrician, in the same manner as a journeyman electrician would train an apprentice. I will let you in on all the tricks of the trade as I know them. The emphasis is on the use of the proper materials and tools for the best results. At every opportunity I outline the simplest, most direct methods of doing the work.

With this in mind, I can give you the following advice: Carefully follow directions, listen closely to the "journeyman," do things right, and *check your work*. With proper tools, instructions, and care you cannot fail to do electrical wiring properly. Wiring can be fun and satisfying, but it is also hard work. When you have completed your work, you will have the satisfaction of having saved a substantial amount of money not only from your own labor but also on the cost of the materials.

James L. Kittle

HOME—ELECTRICAL REPAIR AND MAINTENANCE

CHAPTER 1

Basic Electricity

We are all familiar with what electricity *does,* but not everyone is familiar with what electricity *is*. **Electricity** is defined as the movement of electrons, designated by convention as a minus sign. At one time it was thought that electricity (electrons) moved or flowed from the positive pole to the negative pole, much like the terminals of a battery or dry cell. This theory has now been discarded in favor of the theory that electrons move from the negative to the positive pole.

HOW ELECTRICITY IS GENERATED

Electricity is generated by electric generators similar in appearance to an electric motor. Figure 1-1 illustrates one of the basic rules of electrical generation, the **right-hand rule.** As shown, a conductor with a voltmeter connected across its ends is moved through a magnetic flux field between the poles of a horseshoe magnet in such a way that the conductor is perpendicular to the lines of force. As the wire cuts through these lines of force, the voltmeter will register voltage of a definitive polarity. This basic method of electrical generation can be set up at home as a practical demonstration.

Consider a single conductor on the armature of a simple two-pole generator and apply the principles of electromagnetic induction for various positions of the conductor as it rotates in a magnetic field. (Figure 1-2 illustrates this concept.) The armature conductor (the small circle in Fig. 1-2*a*, taken as 0°) is moving parallel to the magnetic flux lines. Since no lines are being cut, no voltage is induced in the conductor, and no current flows.

As the armature rotates (Fig. 1-2*b*), the conductor begins to cut the magnetic lines (flux) which induces a voltage and current. Application of the right-hand rule shows that current flows into the page, indicated by the ⊕ symbol, and away from the reader.

When the conductor reaches the 90° point (Fig. 1-2*c*), it is cutting the flux at right angles and hence at a maximum rate. The current is still flowing into the page, and it is at this point that maximum voltage is induced. As the conductor continues around, the induced voltage decreases until the conductor again moves parallel to the magnetic flux lines at 180°; then no voltage or current is induced.

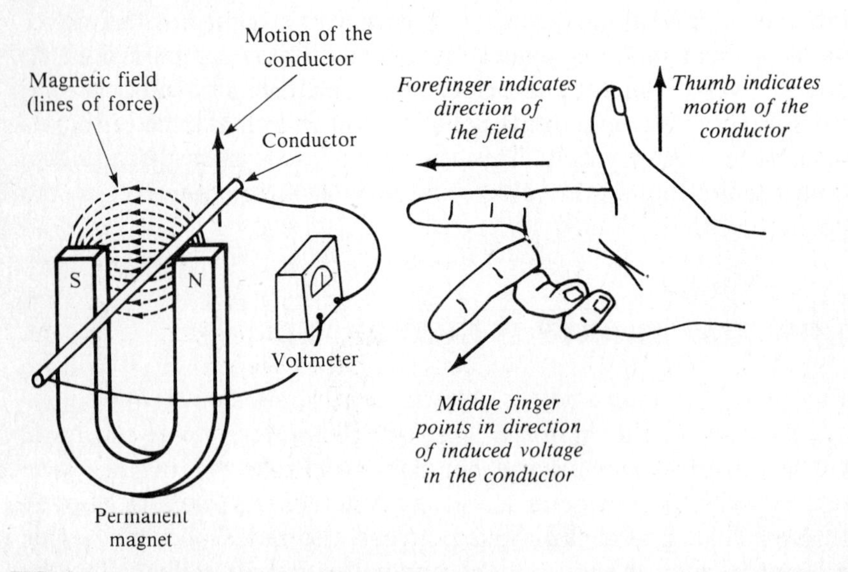

FIG. 1-1 The right-hand rule of electrical generation. (Redrawn from McPartland and Novak, *Practical Electricity*, McGraw-Hill, 1964.)

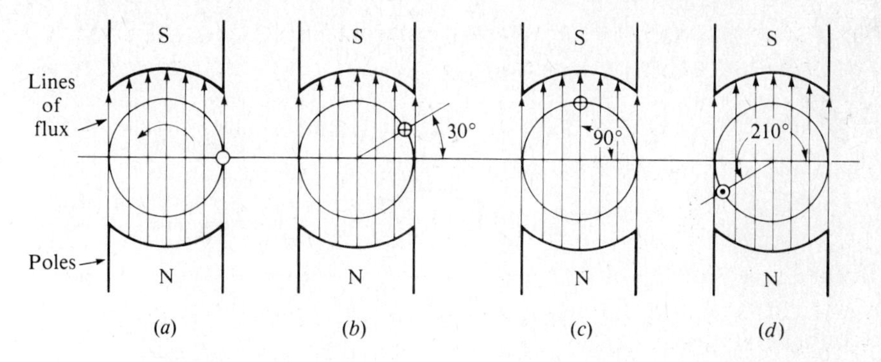

FIG. 1-2 The rotation of an armature through the lines of a magnetic field induces voltage and current. (Redrawn from McPartland and Novak, *Practical Electricity*, McGraw-Hill, 1964.)

As the degree of rotation passes the 180° point (Fig. 1-2*d*), the conductor is again cutting lines of flux; however, this time it is from left to right. Using the right-hand rule, we see that the current now flows *out* of the page and toward the reader, indicated by the ⊙ symbol. In this case, induced voltage reaches a maximum at 270° and then decreases to zero at 360° (or 0°), at which time a new cycle begins.

The production of voltage by relative motion between a current-carrying conductor and a magnetic field is called *electromagnetic induction*. The voltage produced in the conductor is called *induced voltage*. If the wire is closed, that is, complete as a loop, current will flow. This current is called *induced current*.

The generator in generating plants is huge and much more complicated. Numerous turns of wire make up the rotating coils and other parts of the system. The generator is designed to rotate at a high speed and cut all the lines of force, generating a high voltage that alternates or reverses direction (see Fig. 1-3). This produces what is known as *60-cycle current* (60-Hz). The current reverses direction once each revolution and then turns back again before the next revolution, or cycle. Thus, there are 120 reversals of current per second.

Variation in the value of generated voltage is shown when the curve is plotted from the instantaneous values during one complete cycle of the two-pole generator; the result is the familiar sine wave associated with alternating current. All references to alternating half cycles have the same shape, but successive ones differ in polarity. Those above the line are commonly referred to as positive

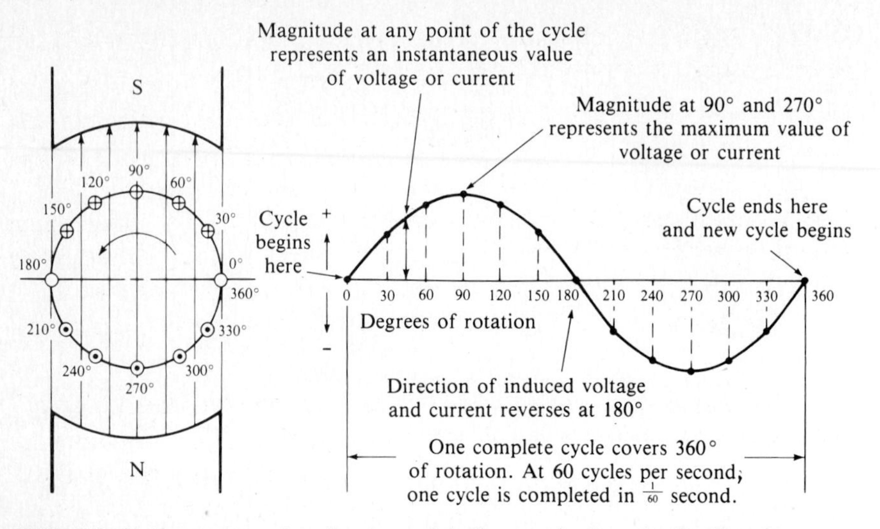

FIG. 1-3 Voltage values plotted on a straight-line graph show the familiar sine curve configuration. (Redrawn from McPartland and Novak, Practical Electricity, McGraw-Hill, 1964.)

alternations, and those below the line as negative alternations.

During the alternations of an alternating voltage or current, the value at any given time is called the *instantaneous* value, and the greatest value reached in each half cycle is called the *maximum* value.

The only current available to households is **alternating current (ac). Direct current (dc)** is not available except in some downtown sections of large cities. Batteries used for such appliances as radios are dc and must be connected properly to the equipment.

TERMS YOU'LL NEED TO KNOW

Many of the terms you'll encounter will seem familiar to you; however, they will acquire new meaning when used in the field of electricity. For simplicity, let's compare electric wiring systems with water systems to explain how electricity works. The following list gives analogies to water systems to better explain how electricity works.

Ampere (A): A unit of electric current that is equal to a flow of one coulomb per second; also abbreviated "amp."

Coulomb (C): A unit of electric charge equal to the quantity transferred by a current of one ampere per second.

Volt (V): Like water flowing through a faucet, electric power is under pressure. The pressure at which electricity is pushed through a wire is measured in volts. For example, the extremely low voltage from a radio or flashlight battery (1½ V) is so slight that it cannot be felt. This is analogous to the force of the gentle spray of a garden watering can. However, the force coming from a high-voltage live wire is comparable to the tremendous pressure of a fire hose.

Watt (W): A unit of power, that equals the work done at a given rate. Watts can be calculated with this equation: volts × amperes = watts. Since they both measure the rate of work, the watt is analogous to the horsepower developed by an engine at a given instant. This is not the same as measuring how much work is being done over a period of time. Taking the wattage drawn by a given appliance or fixture and multiplying it by the time the appliance has been operating, we can arrive at the total amount of electric power used in that period, called **watthours** (Wh). For example, 100 W × 10 = 1000 Wh. The electricity used is recorded by a watthour meter, usually mounted on the rear of your house. The 1000 Wh in the equation is recorded on the meter as 1 kilowatthour (kWh), which costs 5¢–10¢. (The prefix *kilo* comes from a Greek word meaning thousand; hence kilowatt means 1000 W.)

Resistance (R): A wire of certain length and diameter will allow only so much electricity (current) to flow through it, just as a pipe or tube will allow only so much water to flow through. If you try to water your lawn with a very narrow hose (¼-inch outside diameter), you will never finish, no matter what the water pressure is.

Load: The total amount of current drawn or consumed by every appliance, motor, or fixture in a given system. To correctly design a system, we need to spread the load evenly among the circuits. For instance, a light bulb uses 25–200 W, whereas an electric iron or toaster may use up to 1100 W. Although fuses or circuit breakers restrict the amount of load that the home wiring may safely handle, it is advisable not to plug the iron into the same receptacle as the toaster oven. If you do, the fuse may blow or the circuit breaker may trip, to prevent the load from becoming too great for the **capacity** of the wire (otherwise known as its *ampacity*).

Fuse: An "overcurrent device." A fuse acts to stop the flow of current through a circuit when the ampacity rating of the wire that the fuse protects is exceeded. Previously, all plug fuses (the screw-in type, known as an **Edison-base fuse**) would fit into the same fuse holder regardless of their rating. Thus a circuit protected by a 15-A fuse *could* be fused with a 30-A fuse. Clearly this defeated the purpose of the fuse, so a special type of fuse with an adapter, called a **Type S fuse,** was developed. This fuse made it almost impossible to install larger fuses where smaller-rated fuses were required.

Circuit Breaker (breaker): An "overcurrent device." Like a fuse, it acts to stop the flow of current when the breaker's rating is exceeded. Although it costs more, the advantage of a breaker over a fuse is that it can be reset instead of being replaced. If it fails and needs to be replaced, you may be tempted to replace it with a larger size, but do *not* do this. Incorrect replacement seldom occurs because a professional usually handles the job.

Service Entrance Equipment: This equipment is the first part of the wiring system, encountered after the cable has been run into the basement or other location. Included are the *main* breakers. These are used to shut off all the electricity to the building. Then the only hot (live) wires will be those at the top side of these main breakers. More about this in Chapter 5. This equipment will usually house all the *branch* circuit breakers for the building. An exception may be the case in which a breaker panel has been added to handle an additional load.

Ground: For our purposes, the **ground** refers to an underground cold-water pipe connected to a driven pipe or rod. The term **to ground** means to connect something to the ground. Every electric system must have certain current-carrying and non-current-carrying parts, with wires grounded to prevent shocks and fires. (This is an extremely important safety procedure.) Refer to the NEC for complete details and specifications.

Conductor: Any material (usually metal) that will allow electricity to flow through it or along it. Some examples are copper wire, lead pipes, all metals, and you—be careful what you touch!

Insulator: Any material that will *not* allow electricity to flow through it or along it. Some examples are glass, plastic, dry wood, dry cloth, ceramic, and air.

TRANSFORMER: An electric device that converts ac voltage to either a higher or lower voltage. If a transformer converts to a higher voltage, it's called a *step-up* transformer; if it converts to a lower voltage, it's called a *step-down* transformer. A very noticeable type of transformer is the kind found on utility poles near your home. Another type is the familiar doorbell transformer which you may have in your basement or utility room. This kind is often about 2 inches square and is mounted on a metal cover of a junction box or on the side of the box.

ELECTRICAL SYMBOLS AND JARGON

Throughout this book it will be necessary to illustrate correct wiring connections. This information is presented through line drawings using symbols for devices and appliances. Some of these symbols may be familiar to you while others may be new. Therefore, all symbols are explained and illustrated (see Fig. 1-4). Some words used in the trade by journeymen electricians have meanings other than those used in everyday speech. A few of these words will be helpful to know:

Break To disconnect a circuit from the power supply by a switch or other means, such as a control device.

Make To connect a circuit to the power supply by a switch or other means, such as a control device.

Blow A fuse *blows* to break a circuit before the wires get too hot. (It is doing what it was designed to do.) The fuse used for a replacement must be identical in rating to the blown fuse. On a sustained overload, the link across the window of the fuse will usually be melted and just fall away to show a clear window.

Trip A circuit breaker *trips* to break a circuit before the wires melt. The breaker may be reset any number of times.

Hot This word always means live, as a live wire carrying current. Generally the hot wire is the ungrounded black or red wire; sometimes the white wire may also be hot.

Dead This word always signifies a wire carrying no current. Always check for voltage before touching any wire.

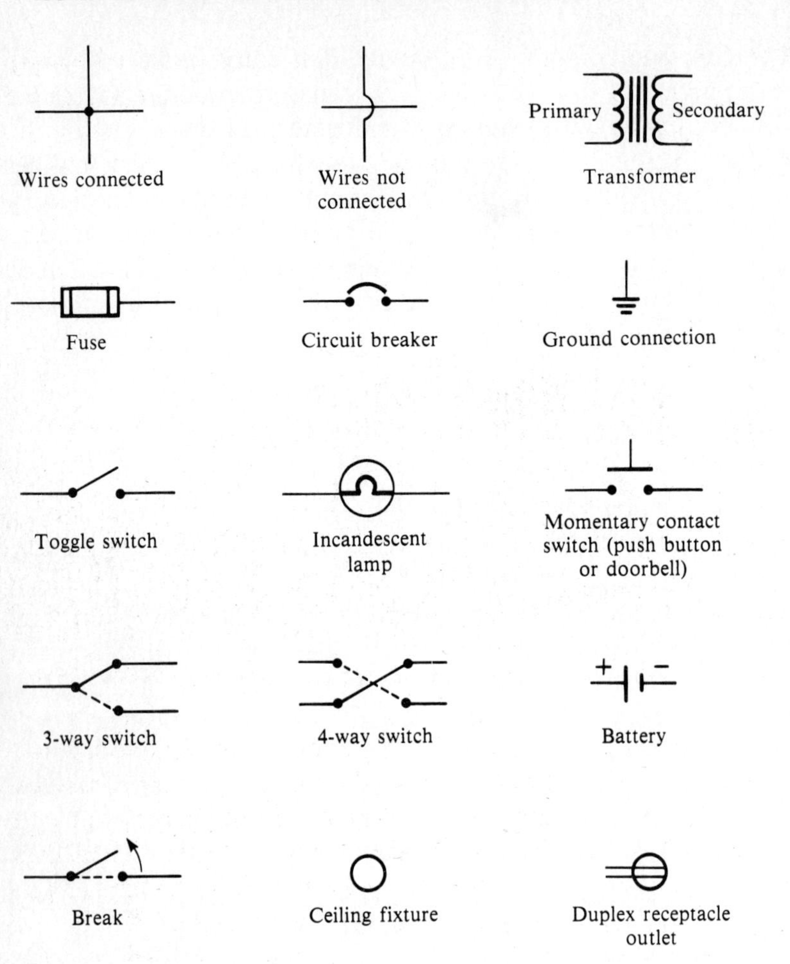

FIG. 1-4 Electrical symbols.

OUR SOURCE OF SUPPLY OF ELECTRICITY

The electricity we use daily is supplied by power generating stations housing large electric generators. These generators are turned by steam turbines at high speed. The turbines are supplied

FIG. 1-5 Overall view of coal-fired generating plant. (*Courtesy of Detroit Edison.*)

with steam from boilers fired by coal, oil, or nuclear sources. Another source of power is water (hydroelectric power). This power is provided by damming rivers to provide a height of water. The power plant is built below the dam and the water falling from a great height then turns water turbines which in turn spin the generators. (Boulder Dam is an example.) When a coil of wire is turned rapidly within the magnetic field of a magnet, electric current is generated. This is the principle of the generation of electricity discussed earlier (see Fig. 1-3). The way in which the wire cuts through the magnetic field causes a variation in voltage from zero to maximum and maximum to zero; it then continues past zero to the maximum in the *other* direction and back to zero. This generates the alternating current that we use in our homes.

Electricity is generated at 12,200 V at the power plant (see Fig. 1-5). Outside the power plant, the voltage is increased to 120,000 and sometimes 325,000 V, depending on the distance of the transmission lines (see Fig. 1-6). When the electricity reaches its destination, such as a city, the high voltage goes to a substation. The substation then reduces the voltage to either 4800 or 13,200 V, depending on local conditions. The method of raising or lowering the voltage is through the use of transformers (see Fig. 1-7).

FIG. 1-6 Steel transmission tower carrying high-tension (voltage) lines away from a power plant.

FIG. 1-7 View of transformer at a power plant.

A transformer consists of two coils of insulated wire, each isolated from the other by its own insulation (see Fig. 1-8). These two coils are wound around a core consisting of thin sheets of soft silicon iron (laminations). These laminations are stacked (layered) to form a hollow shape much like the letter O with square corners. The top of the O is removable, to allow the prewound coils to be

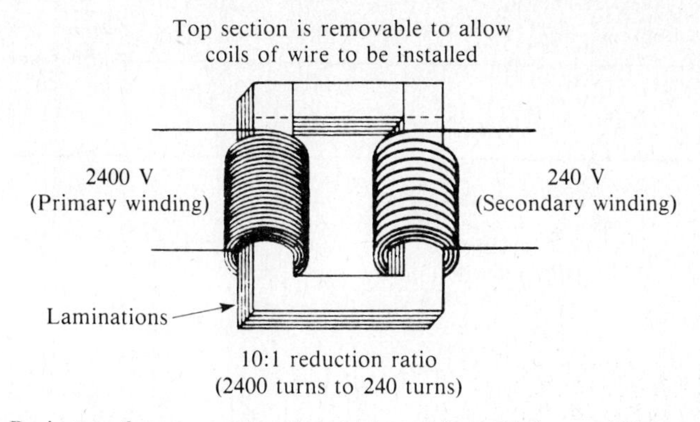

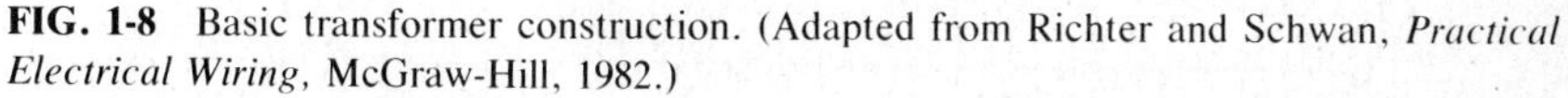

FIG. 1-8 Basic transformer construction. (Adapted from Richter and Schwan, *Practical Electrical Wiring*, McGraw-Hill, 1982.)

slipped onto the two arms. The ends of the two coils extend out from the transformer case for connection.

The *primary* coil has, for example, 2400 turns of wire. The other coil, the *secondary*, has only 240 turns. Because both coils are on the laminations, or core, the primary coil will *induce* a voltage in the secondary. If the voltage to the primary is, for example, 2400 V, then the voltage induced in the secondary will be 240 V. The ratio of the turns is the same as the ratio of the two voltages, 10 to 1.

> ***Note:*** The number of turns does not have to match the quantity of voltage.

Transformers can either raise or lower the voltage depending on the circumstances of use. The home has one or more transformers, in the doorbell, furnace-ac.

> **CAUTION:** Treat the low voltage with respect, because a defective transformer *can* leak the high voltage through into the low-voltage wires.

The top wires on the pole at the rear of your yard will be either 2400 or 4800 V. In either direction from the pole nearest your home run lower wires. Usually every block has one transformer for its needs. This transformer takes the 2400-V power from the top wires, reduces it to 120/240 V, and feeds it into the lower wires, which in turn supply your home, as shown in Fig. 1-9. The trans-

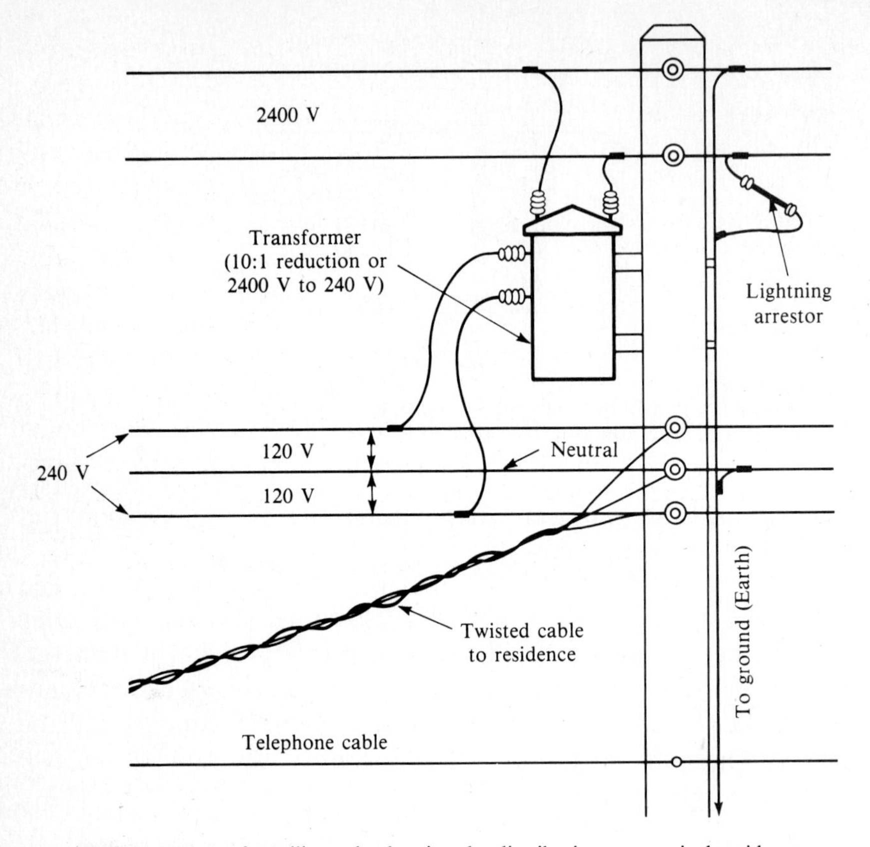

FIG. 1-9 View of a utility pole showing the distribution to a typical residence.

former looks like a tall garbage can mounted on the utility pole with wires coming from it to the horizontal wires running from pole to pole. If you have underground service, the transformer will be a 3 × 3 × 2 ft metal cabinet mounted at ground level on a concrete pad somewhere in your area.

The normal power supplied to homes, as already mentioned, is 120/240 V (sometimes called 115/230 V). The lower voltage (120 V) is used for lighting and small portable items, such as mixers, irons, power tools, and vacuum cleaners. The higher voltage (240 V) supplies major appliances, such as electric ranges, electric clothes dryers, and electric water heaters. Central air conditioning also requires the higher voltage.

Go outside your home and inspect the connections from the utility to your home. If the service is overhead from a utility pole to the dwelling, it is called a *service drop* (see Fig. 1-10). The wire is "dropped" from the high pole to your house connection. Note the messenger cable attached to the building wall—this is the neutral. The building neutral connects to the messenger cable in the center near the service entrance cable. The connector is unwrapped. If the dwelling is a ranch type, a piece of **conduit** (metal pipe holding the wires) may extend up through the roof a few feet, the cable from the pole will be anchored below the top of the conduit service head (a fitting at the top of the conduit to prevent rain from entering), and the wires of the cable will be connected to those coming out of the service head. At about eye level the utility meter will be mounted on its base (sometimes called a meter socket). Continuing down from the meter base will be either conduit or flexible cable; the trade name is *service entrance cable* (see Fig. 1-11). This is the route of power into the basement. If power is supplied underground (called *service lateral*), it will come into the bottom of the

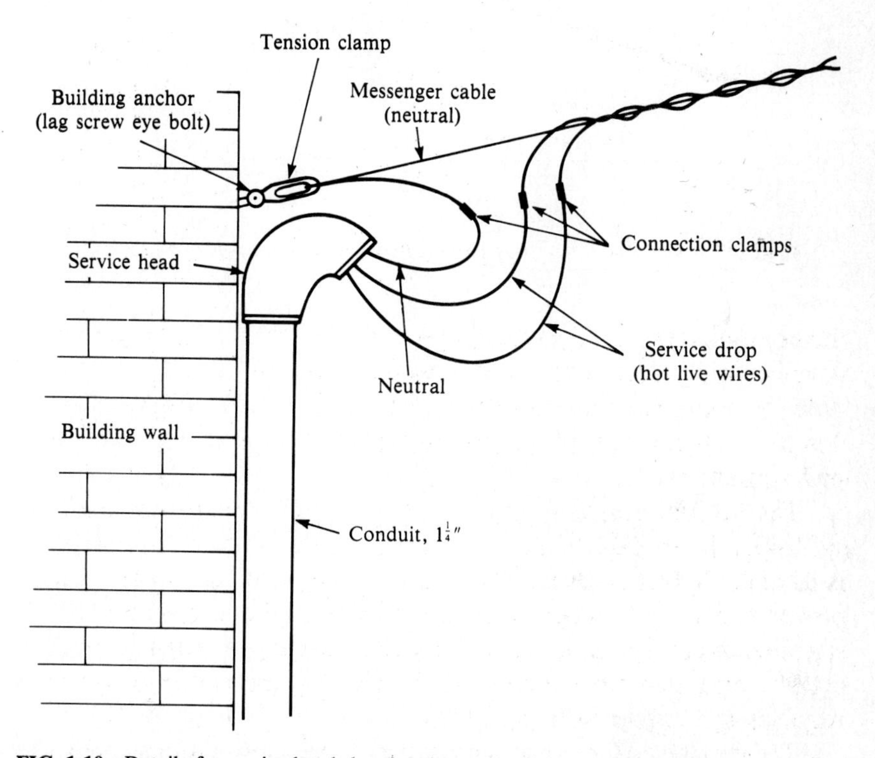

FIG. 1-10 Detail of a service head showing the connection to the service drop at a dwelling from the transformer in Fig. 1-9.

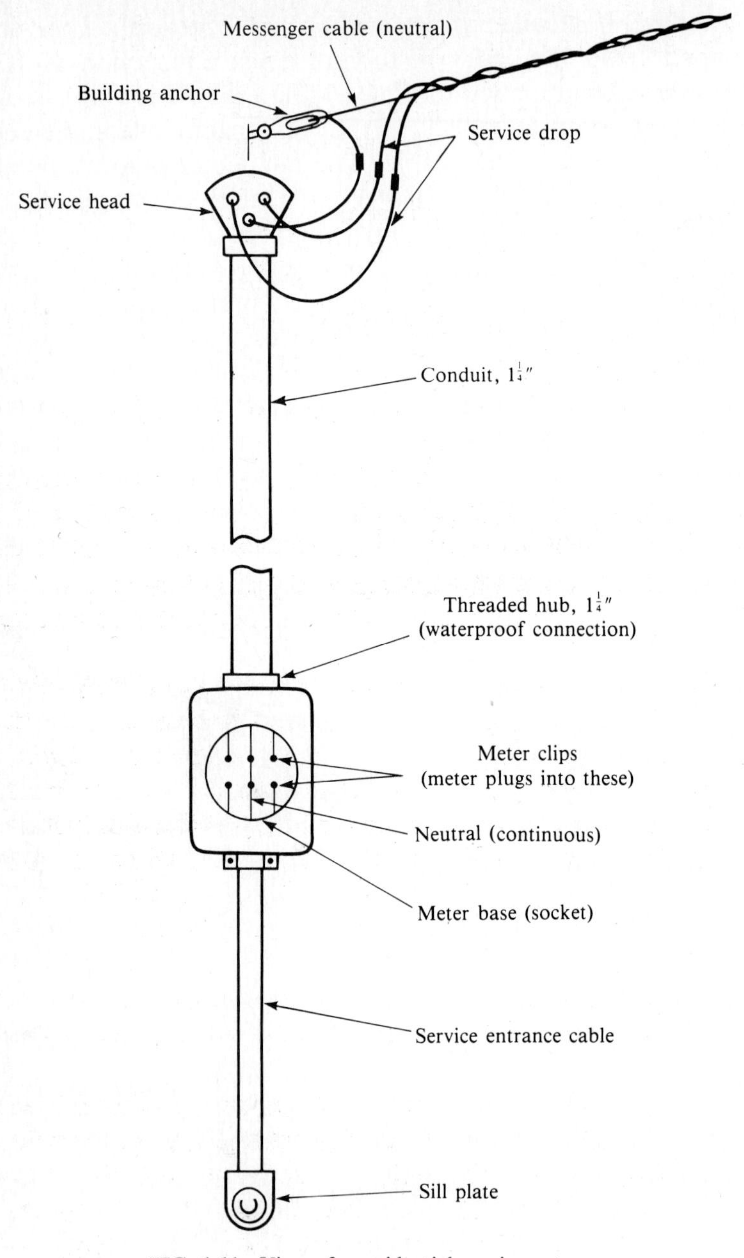

FIG. 1-11 View of a residential service entrance.

meter base through conduit, go through the meter, and then leave through the bottom of the base to enter the basement.

> ***Note:*** Homes without basements usually will have the electric panel in the utility room on the first floor.

Modern service drops consist of three wires: two insulated *hot* wires and one *neutral* wire (see Fig. 1-12). The neutral wire is made of aluminum with a high-tensile, galvanized steel core to give it strength. The hot wires are spiraled around this cable, called the *messenger conductor,* to form the assembly.

> **CAUTION:** Touching hot wires will give you a serious shock.

Now go back inside to the basement, or wherever the cable enters the house, and find the service entrance equipment (see Fig. 1-13). The panel is usually located very close to where the cable entered. This is a requirement of the NEC to protect all wiring inside the building from grounds or short circuits that could cause a fire.

Refer to Figs. 1-14, 1-15, and 1-16 as you read the following section. In warmer areas the main circuit breaker may be in the same outdoor cabinet as the meter. The main circuit breaker, the branch circuit breakers, and fuses are known as *overcurrent devices*. All overcurrent devices are designed to allow only a predetermined number of amperes to flow through the wires which these devices protect. Standard capacities include 15, 20, 25, and 30 A. Wires carrying more current than their rated capacity (**ampacity**) will overheat enough to start a fire if there are combustible materials close by. You should never replace a blown fuse with one of larger capacity rating because the wire could overheat and cause a fire.

Just as a large pipe can carry more gallons of water per minute than a small one, the larger the cross-sectional area of the wire, the more amperes it can carry safely. Accordingly, each size of wire is classified by its ampacity rating. For example No. 14 copper wire will safely carry 15 A of current; No. 12 copper wire will safely carry 20 A. Aluminum wire is still being used for home wiring, but its use is discouraged because of its many drawbacks. Aside from its brittleness, larger diameters must be used. For example, No. 12

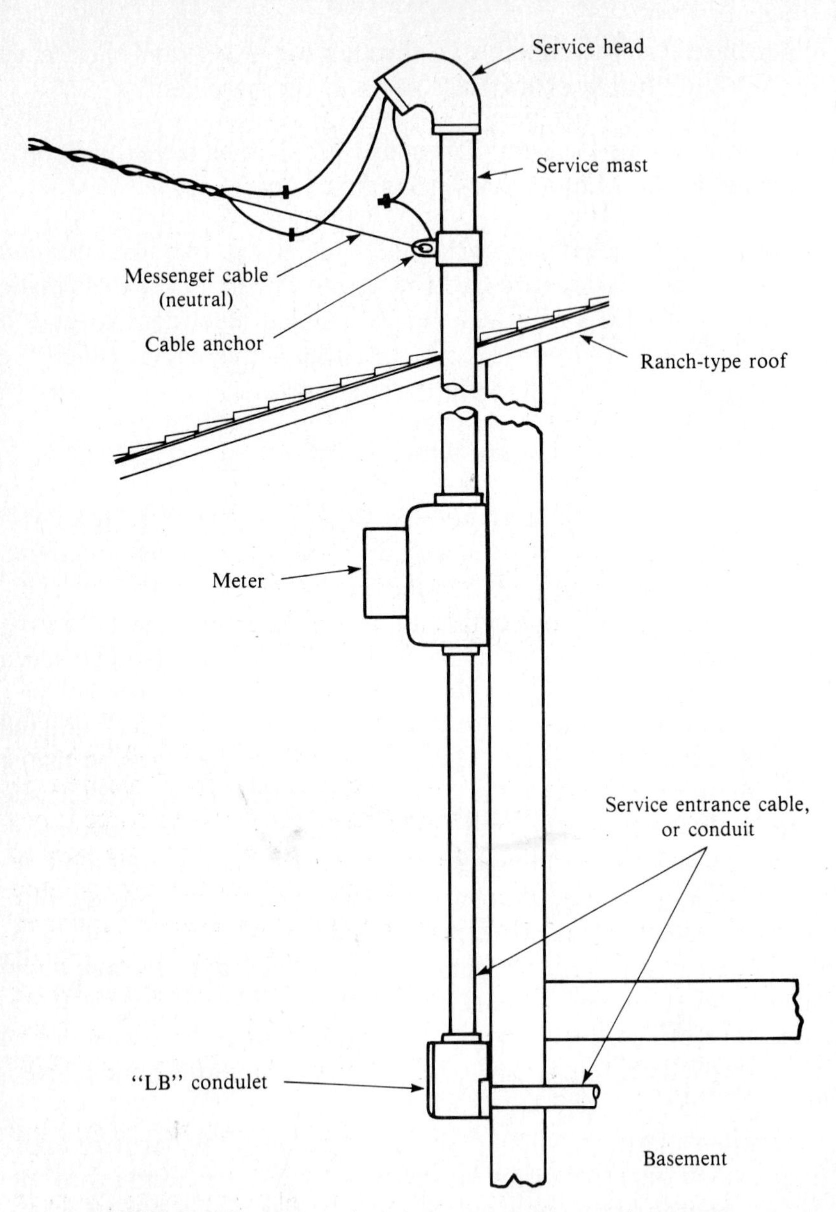

FIG. 1-12 View of a service entrance mast with a ranch-type roof.

aluminum wire will only carry 15 A, and No. 10 aluminum wire will only carry 20 A safely. These ampacity ratings are based not only on wire size but also on insulation type and whether the wire is single or bundled with other wires in a cable or conduit.

Study Figs. 1-9, 1-10, 1-11, and 1-12. These diagrams illustrate

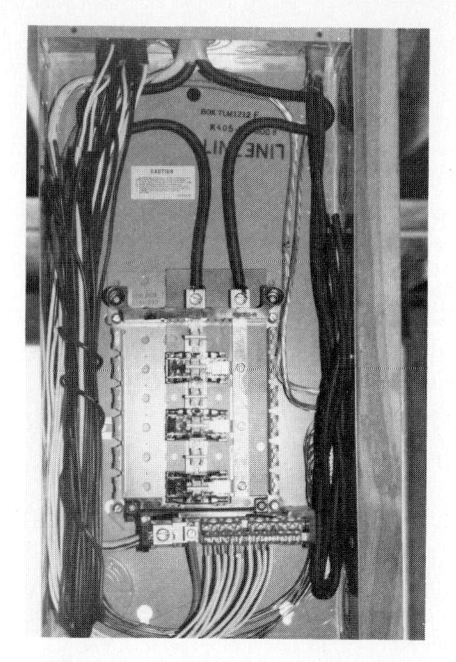

FIG. 1-13 Circuit breaker load center.

how the power is brought to your house. The service drop consists of three wires: two hot wires and one neutral wire. This neutral wire is known as *ACSR* (aluminum conductor steel-reinforced) *cable*. This aluminum cable with a steel core has the strength to withstand high winds and ice. The two hot wires are twisted around the steel cable to form the *messenger cable*. It is furnished in reels by the manufacturer. If the service mast head extends through the roof of a ranch-type building, the cable at the house end is anchored near the top of the mast (see Fig. 1-12). (This head is the gooseneck fitting at the top of the mast. It keeps rain water from entering the conduit at this point.)

If the gooseneck is below the roof and under the eaves, the service drop is anchored to the house frame (see Fig. 1-11). The wires pass through a plastic plate held in place by the gooseneck cover. Each wire has its own hole in the plate to keep them separate. Normally the service drop wires are attached to the mast below the head, so the wires must go up into the service head and down to the meter in order to stop rain and other moisture from entering the conduit. If the service drop must be anchored above the service head, then the wires from the service drop *must* go down, make a loop, and go back up to the service head, also to

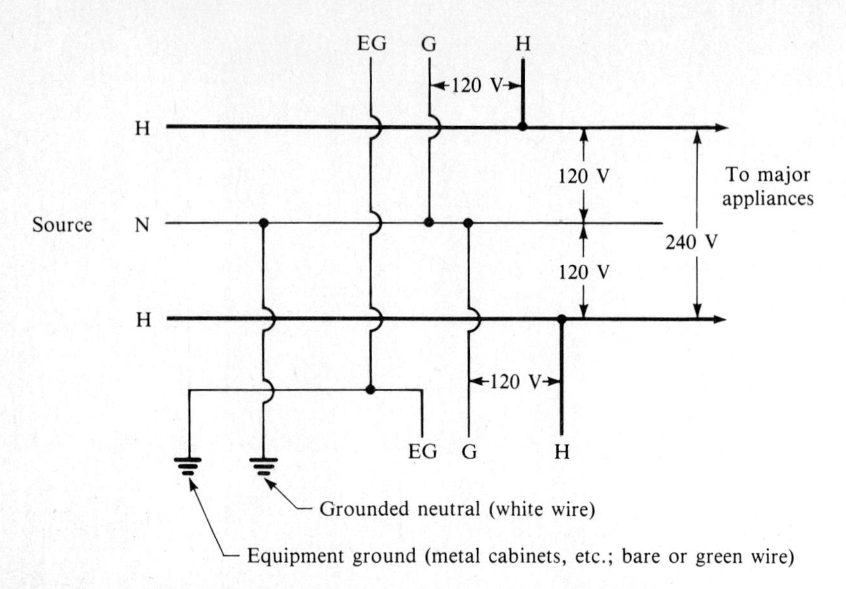

FIG. 1-14 Typical wiring diagram for a service entrance panel. The power comes in at the left. Voltage between the H and H wires is 240 V (shown at far right). Voltage between either H and N is 120 V (shown at top center and bottom center). The neutral is grounded to a cold-water pipe or other suitable ground. EG, also grounded to the water pipe, grounds all metallic non-current-carrying parts, metal cabinets, conduit, and related parts. Note that the neutral (N) becomes the grounded wire (G) when power is supplied at 120 V. EG is the bare ground wire in Romex.

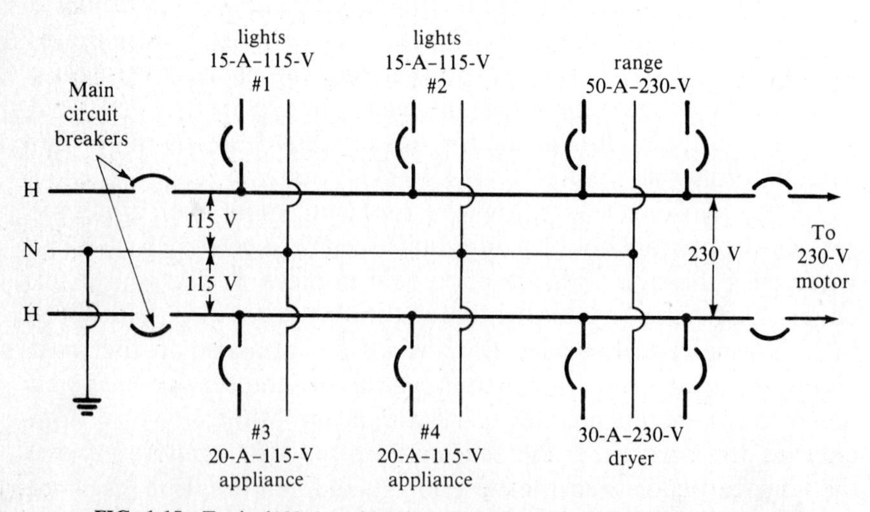

FIG. 1-15 Typical100-A residential wiring diagram with circuit breakers.

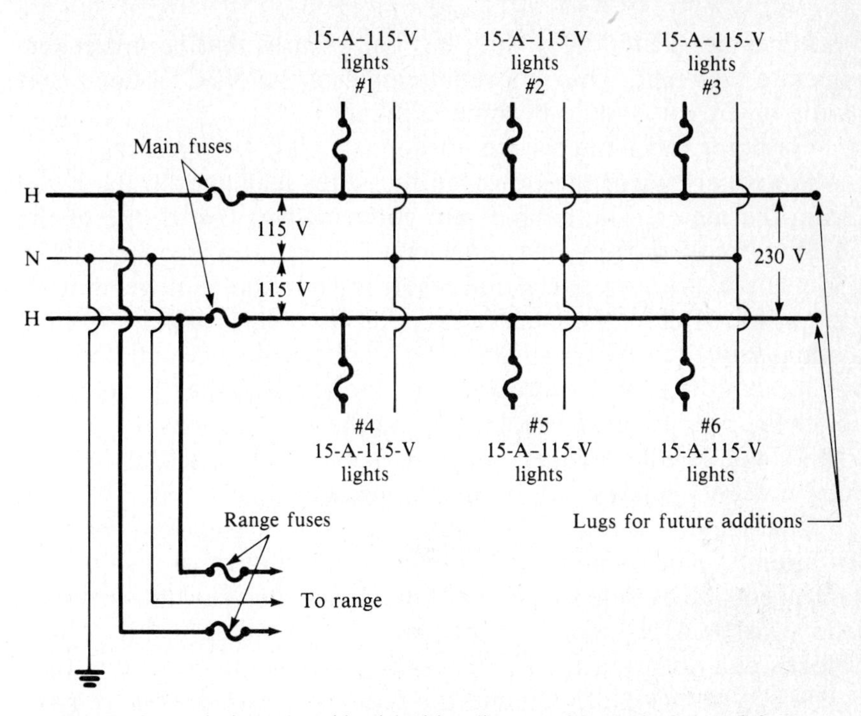

FIG. 1-16 Typical 100-A residential wiring diagram with standard Type S fuses.

prevent moisture from entering. Many installations do not use conduit below the service head but use service entrance cable from the head to the meter. Although it is approved, this method does not look as neat and professional as conduit.

On ranch-type houses the mast is mandatory because the NEC requires that the service drop be at least 10 ft above the finished grade. Although the majority of installations use cable either above or below the meter and into the basement and to the service entrance equipment, cable may not be used in this case. Conduit must be used to provide an anchor for the service drop messenger wire (the actual neutral). Normally the service drop is anchored below the service head to prevent water from traveling down into the head.

SERVICE ENTRANCE EQUIPMENT

Generally service entrance equipment, whether in the basement, utility room, or some other location, will have the main circuit breaker at the top of the panel. This looks like two large toggle switches connected by a bar. This bar ensures that *both* breakers operate as a unit. This is a requirement of the NEC, a document with which you should become familiar.

Tripping this main breaker disconnects (kills) all power (240 V) to the panel except the power in the wires leading into the panel from the meter. The handles are connected so that if one of the double breakers trips, the other one will trip too (see Fig. 1-17). Some breakers are vertical and others are horizontal, depending on the maker. Below the main breaker are two vertical rows of individual *branch circuit breakers*. The number will vary with the size of the building. Those breakers for the range, water heater, and clothes dryer are also double because these appliances use 240 V. The single-handle breakers serve 120-V circuits, such as lighting and wall receptacles for portable appliances.

Service entrance panels that use fuses are designed somewhat differently. The usual type has two plastic removable fuse blocks arranged side by side at the top of the panel. One is embossed with the word MAIN, and the other with the word RANGE. These blocks pull out from their respective recesses and carry the fuses and have prongs that plug into the bottom of each recess to make contact with power. The range and main blocks are not interchangeable in their respective recesses. As long as the fuse block is away from the panel itself, there is absolutely no danger of shock. When you change or replace the fuses, be sure to use the same rating of fuse, *never* larger.

Arranged below these pull-out blocks are the branch circuit fuses. These are usually 15- or 20-A ratings, the NEC limit. One way of recognizing a 15-A fuse is by its hexagonal window shape. All other rating sizes have round windows. The 25- and 30-A fuses are available, but it is not advisable to use these sizes unless the wires in these circuits have the ampacity necessary to handle these ratings. Note that these old-style Edison-base fuses are no longer permitted in new or rewired systems. The newer Type S fuse has taken the place of the Edison-base fuse. A Type S fuse is installed by first inserting a permanent adapter into the fuse socket that will accept only the correct size Fustat (the trademark of Bussman Fuse Company, the developer of this type of fuse). This makes inserting the wrong size fuse virtually impossible. In addition, pennies and foil can no longer be used to bridge a blown fuse.

FIG. 1-17 Plug-in circuit breakers.

Any system which is properly designed, balanced, and installed will have very few problems with breakers tripping or fuses blowing. If problems occur, they could be caused by appliances and power tools, not by the system itself. Overloaded circuits will occur when there are too many appliances or there is too great a current draw for the circuit design.

Most homes have similar wiring systems differing only in the system capacity and size of service equipment. The number of circuits depends on how "all electric" the home is designed. The typical house may have the following circuits:

1 Range circuit	240 V	50 A
1 Furnace-ac circuit	240 V	30 A
1 Dryer circuit	240 V	30 A
1 Laundry circuit	120 V	20 A
1 Disposer circuit	120 V	15 A
1 Garage door circuit	120 V	15 A
4 Lighting circuits	120 V	15 A
2 Small-appliance circuits	120 V	20 A

In certain areas, electric water heaters are on a separate meter at a special rate and are not included in the total connected load (the actual items using current plus the estimated normal load expected). In the example above, the panel provides spaces for twenty-four $\frac{1}{2}$-inch-wide or twelve 1-inch-wide circuit breakers.

Recall that 240-V breakers are two-pole and therefore take up two spaces, regardless of whether they are $\frac{1}{2}$ or 1 inch wide, and allowance must be made for their size. The 240-V breakers are commonly 1 inch wide and so use four of the $\frac{1}{2}$-inch spaces. Not all spaces on the panel are used in the average house; these can be saved for future expansion.

Most panelboards are supplied with the main breaker already installed. In condominiums and apartments, the main breaker is incorporated into the meter housing on the outside of the building. Breakers for the branch circuits are purchased separately because they are selected to meet the requirements of each individual circuit. Refer to the list above for breaker sizes.

PANELBOARD AND CIRCUITS

Numerous runs (lengths) of cable leave the panel to feed various areas of the house. These are either **Romex** or **BX.** Romex is most common and refers to *nonmetallic sheathed cable*, its generic name. It consists of two or three insulated wires plus a bare grounding wire, all covered with a tough outer covering of plastic or fiber. BX has a galvanized-steel covering that is wound and interlocked spirally around the insulated wires inside. A grounding strip of aluminum inside acts as the grounding wire. An insulating bushing (plastic tip) is inserted between the insulated wires and the cut end of the outer metal covering to protect the insulation on the wires from damage. Although Romex is a registered trademark, it is also the generic word used for this type of cable just as Frigidaire is often used for any make of refrigerator. BX is permitted where Romex is prohibited by local ordinance. It is more expensive but has higher resistance to damage.

In most cases, the modern house will have *lighting circuits* that are protected at 15 A and use No. 14 wire. (There are exceptions, however; New York City, for example, requires No. 12 wire.) The kitchen and dining areas will have *small-appliance circuits* protected at 20 A and using No. 12 wire. Other circuits, called *fixed appliance circuits* for range, dryer, furnace-ac, and water heater, use heavier wire and breaker for fuse protection. The range is protected at 50 A and uses No. 8 wire. The other appliances are protected at 30 A and use No. 10 wire. Romex or BX cables leaving the distribution panel for these circuits are larger in physical size. The range and dryer circuits are allowed to use service entrance cable, the same cable used on the service entrance assembly, usually below the meter and into the building. The furnace-ac and the water heater circuits use Romex or BX as in other circuits, only with larger wire sizes. The range and dryer rate a special exception to use the service entrance cable. The NEC makes this exception as long as the service entrance cable originates at the main service entrance panel.

EVALUATING NEEDED ADDITIONS

In general, each circuit will consist of ten or twelve receptacles. By definition, the word *outlet* signifies any point where electricity may be used, such as a ceiling fixture or a receptacle. After you have inspected the service entrance panel and breaker or fuse panel and have calculated the present load and any proposed additional load, you may want to add an auxiliary panel to provide space for more circuits. In considering this extra panel, the size of the main breaker or fuses and the size of the wires or cable should be checked to see whether there is sufficient capacity for the additional load. Chapter 4 explains and illustrates types and sizes of cable and wire and provides you with the correct procedure to calculate the present load on your wiring system. If you have carefully removed the front panel cover, you may be able to judge the size of the wire feeding the entrance panel. The amperage rating of the entrance panel is on the specifications sheet glued to the inside of the panel door. An entrance panel using fuses will be 60 A. An older, larger house may have a larger panel rated at perhaps 100 A.

The modern house will have only circuit breakers, and the panel will usually have space for additional breakers, as previously described. Adding a breaker where there is space is easy. Be sure to buy a suitable breaker for the brand of panel that you have (G.E., Leviton, and so on). The breaker package will include a list of compatible entrance panels which will accept that type of breaker. Chapter 7 explains how to install a new breaker and connect the wiring to it.

Safety, the NEC, and the UL Label

SAFETY

All kinds of work require taking certain safety precautions. The installation of electric wiring further requires the implementation of correct and accurate procedures to ensure the long life of the system. Home mechanical equipment consists of plumbing, heating and cooling, and electric systems. Because these systems operate throughout their useful life, the mechanical equipment wears out, breaks down, leaks, or just plain won't work.

To ensure both long life of the system and safety of people and property, electric wiring *must* be installed correctly. The right wires must attach to the correct terminals, terminal screws must be tightened enough to make a good connection, no bare wires or wire strands must be left exposed (grounding wires excepted), and all precautions must be taken to ensure that other persons will not be harmed by a faulty installation. Wires to terminal screws must be curled around the screws in a clockwise direction a bit more than two-thirds of the way around the screw shank. Extreme care must be taken when working around live wires to avoid serious electric shock.

Most electric devices now come with installation instructions and diagrams. Other precautions, such as marking the maximum bulb size on a fixture, prevent the fixture from overheating. Some fixtures come with a glass-fiber pad to insulate the fixture wires from the fixture heat. This pad *must* always be in place.

Incorrect wiring, such as the interchanging of black and white wires to a receptacle or motor, has serious consequences. Electricity is a useful tool when wiring is carefully installed and maintained and, more important, when persons and property are protected from hazardous conditions. Respect electricity, and it will serve you well.

GROUNDING

The NEC is greatly concerned with the grounding of equipment and wiring. Grounding is important and must be maintained for all metal parts of the system that do not *normally* carry current (current does flow through these parts when there is a short circuit to any metal equipment part, such as a box or conduit). Grounding prevents a serious or fatal shock caused by touching a piece of equipment with a defect that causes the case or cover to become live. Because shock tends to cause a person to hang on to the live part, quick action is needed to remove the wire from the person.

> **CAUTION:** Do not touch either the person or the live wire! If you do, you will be shocked also.

If possible, shut off the power. If you cannot do that, try to pry the wire away from the person, using a board or other insulating material. Once the person is free of the live wire, start CPR (cardiopulmonary resuscitation) and have someone else call for medical aid. To repeat, quick action is imperative! This is a very serious situation.

Most modern houses have 120/240-V service running into the house. Three wires enter the service entrance panel at the top. If they enter from the service entrance cable, there will be one black, one red, and one bare wire, which is the neutral. The bare wire in the cable is wrapped spirally around the two insulated wires and must be unwrapped and formed into a wire shape. The three wires coming to the panel from the conduit may all be black, with the neutral wrapped in white tape to identify it as the neutral.

The panel will have two terminal bars. One is called the neutral bar, and the other is called the equipment ground bar. At the top of the panel are the two terminals for the two hot wires. These two wires will connect to either a double circuit breaker or a fuse holder. Even though the breaker is tripped or the fuse block is removed, these two terminals will remain hot. Be very careful not to touch them. In many panels the neutral bar is some distance from these hot terminals, either alongside the circuit breakers for the branch circuits or at the bottom of the panel. The neutral bar accepts the neutral wire from the entrance cable. The grounding wire to the system ground, such as a buried cold-water pipe, is also connected to this neutral bar. Since these two ground wires are large in size, there are large holes in the bar just for these two wires. Both the neutral bar and the equipment grounding bar are connected to the metal cabinet by metal straps.

All the white wires in the cables that run to the various circuits are connected to the *neutral* bar, whereas the black or red wires are connected to the terminals on the breakers or fuse holders. The bare ground wires in the Romex or BX are connected to the *equipment* grounding bar. Be sure to connect the wires going to these two bars correctly. At the water meter, there must be a jumper cable connected to both sides of the meter but a short distance away from the meter itself. This jumper cable is installed to give continuity to the ground system in case the meter is removed for any reason. The ground wire and the jumper wire are usually No. 6 stranded wire stapled to wood supports in the basement.

This continuous grounding system maintains an earth connection to all appliances and power tools provided with a three-wire cord and plug and to three-wire receptacles that are properly wired. When an appliance or power tool develops a defect, such as a bare wire inside touching the casing, the casing will be hot and could cause a severe shock. If this appliance is then plugged into a grounded receptacle, the fuse will blow and render the appliance harmless.

CAUTION: Do *not* use the appliance until it is repaired.

The latest safeguard developed is the **ground fault circuit interrupter** (GFCI), which acts within a fraction of a second to disconnect the circuit even though the appliance does not have a three-wire grounding cord and plug. According to the NEC, the GFCI is mandatory in bathrooms, patios, and garages.

It is important at all times to be aware of what you are doing. You should follow recommended methods and procedures to ensure a safe and correct installation. Be sure to wear protective goggles when you are pounding, chipping, and even looking up to do work overhead.

NATIONAL ELECTRICAL CODE

Earlier, brief references to the NEC have been made with no full explanation. The NEC is a compilation of standardized methods of the installation of electric wiring and electric devices. The code is published by the National Fire Protection Association, which also sets standards for fire protection and fire safety. The NEC in its entirety is adopted by most governing bodies such as cities, townships, villages, counties, and states, and made a part of their laws and ordinances. Thus, the code may be enforced by the governing body having local jurisdiction.

Since many fires and lost lives result from faulty wiring and installation procedures, governing bodies retain this control over electric wiring installations. Electric wiring is a very precise trade requiring concentration and double-checking of work done. The proper connection of wires is crucial to safety and correct operation. This trade is not hard to learn, but strict attention must be given to the job at hand; this applies to the do-it-yourselfer as well. Always read all directions and follow them carefully.

At this writing, the complete NEC in paperback costs $10.25; the abridged NEC entitled *Electrical Code for One- and Two-Family Dwellings* costs $8.25; and *The National Electrical Code Handbook,* which consists of the complete NEC plus extensive explanatory material added costs $27.50. All prices are postpaid. The next editions of the 1987 NEC will be out in late 1986. Be sure to write for prices if you decide to buy a copy. All three editions are available from the publisher by direct mail. Libraries usually have copies for circulation and reference, including current and previous editions.

While the NEC is not law as published, when adopted by city or county governments by reference it has the force of law. The NEC promotes safety and adequate wiring methods and materials.

NEC rules promote safe and correct wiring, but they do not

necessarily specify the adequate wiring capacity for the building requirements. Safety of personnel is also of very great concern in the formulation of NEC rules. Fire safety is also important because of rebuilding costs and loss of manufacturing facilities. For every part of an electric system rules are formulated for its use, including installation, types of materials, and methods of wiring. For example, suppose you want to replace a defective duplex *nongrounding* receptacle with a grounding receptacle having the *third round grounding* opening connected to the metal wall box by a wire to the green *hex* grounding screw on the metal supporting yoke of the receptacle. This ground wire must be clamped to the nearest cold-water pipe or, as the NEC states, ". . . shall be permitted to be grounded to a grounded cold water pipe near the equipment." As you can see, the NEC is very strict, and for good reason. A person using an electric power tool grounded through the third prong of its attachment plug will assume that this receptacle is actually grounding the power tool when it may not, because the receptacle itself is not connected to a ground (as required).

The NEC provides methods of calculating the minimum requirements for dwellings that provide only for safety of persons and property from hazardous conditions. Anything used in the wrong manner and not for the purpose intended is hazardous to both persons and property, such as lamp cord used as a wiring material in place of Romex cable or BX cable (cable having a spiral steel jacket and required in areas such as Chicago). Persons who are not familiar with standard wiring practices sometimes run into trouble with their wiring installations. I strongly urge you to purchase either the complete NEC or the *NEC Handbook*. You can also borrow current or earlier editions from your local library.

The NEC for 1984 consists of nine chapters, including these topics: wiring design and protection, wiring methods and materials, equipment for general use, special occupancies, special equipment, special conditions, and communications systems. Each chapter is divided into sections dealing with certain types of equipment or methods.

The NEC provides instructions on how to connect a system to a proper ground, states clearance requirements for the service drop (the height above ground level over private property versus over public roadways), gives instructions on the running of cable such as Romex, and details the methods of fastening to supports. Tables are included that state the ampacity of various wire sizes for single or cabled wires as well as the different insulation materials covering them.

Because the writing style of the NEC is formal, you may have to read carefully to understand the proper procedures. This is one of the many advantages to having the *NEC Handbook,* which offers descriptive explanations and includes diagrams and illustrations for clarity.

Especially noteworthy is Article 300 on wiring methods which is most helpful, as well as Chapter 3. ''Equipment for General Use,'' Chapter 4, includes sections on lighting fixtures (Sec. 410), motors and motor circuits (Sec. 430), and air conditioning (Sec. 440). A comprehensive index is included.

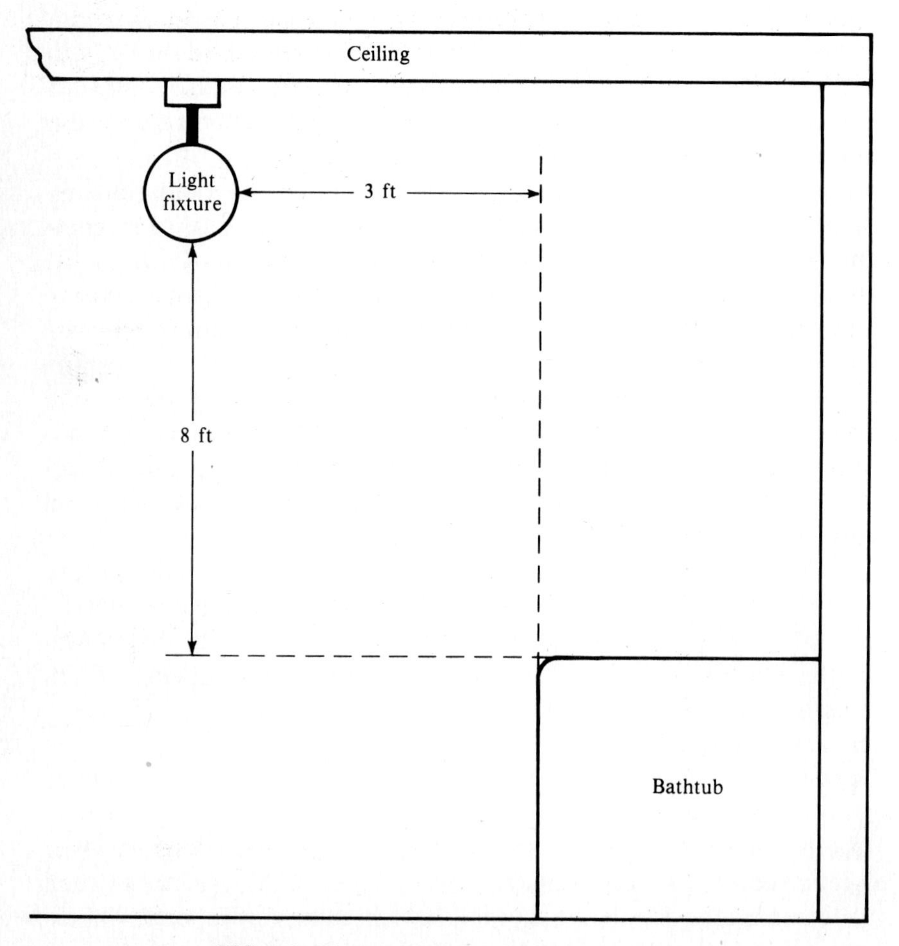

FIG. 2-1 Diagram illustrating NEC Sec. 410-4d.

The NEC is revised every three years to keep abreast of new products, developments, and services and to revise and clarify code wording if necessary. A recent NEC change affects attachment plugs used on vacuums, toasters, power tools, and anything that can be plugged into a receptacle. Plugs are now required to be "dead front" design. In the previous design, the wires and screws on the prong end were exposed. A fiber cover was used to cover these exposed connections. When the plug was inserted into the receptacle without the fiber cover and the receptacle plate was metal, any wire strands contacting the receptacle plate could cause a short circuit, arcing from the stray plug wire strands to the metal plate. In the new design, the wire ends and screw terminals are placed inside the plug body, and this eliminates the problem. Both two- and three-prong grounding plugs are available in all price ranges depending on your needs. Figure 2-1 shows clearances from the bathtub to a hanging fixture. (See also Appendix A.)

The NEC stresses safety frequently because it must always be kept in mind when you are working with electricity. *Always turn the power off when working on wiring*. Be sure to check for power *twice* in the section you are working on. Accidents occur even under the best of circumstances, so use ordinary caution, think about what you are doing, do it right, and check your work carefully. Even experienced electricians are cautious when turning on power to a *new*, untried installation.

When *every* connection is made securely, wired properly, and maintained efficiently, you will have a safe electric installation. This complete system is seen in appliances having a grounding-type attachment plug (three-pronged) inserted into a properly grounded receptacle.

> CAUTION: Never cut off the round grounding prong on an attachment plug. This negates any protection available through the grounding receptacle. The whole grounding system is for your protection. Use it.

THE UL LABEL

The UL label (see Fig. 2-2), the symbol and copyrighted trademark of the Underwriters Laboratories, Inc., which tests and lists electrical appliances, wire, and wiring materials plus the products of manufacturers in other fields, is mentioned in the NEC only as a

FIG. 2-2 The Underwriters Label on the cover of an auxiliary fuse panel.

HOME OWNER AFFIDAVIT

LOCATION________________________________ DATE________________

As the bona fide owner of the above mentioned property which is a single residence, and which is, or will be on completion my place of residence and no part of which is used for rental or commercial purposes nor is now contemplated for such purpose, I hereby make application for an owner's permit to install________________________________ **as listed on the permit application.**

I certify that I am familiar with the provisions of the applicable Ordinance and the rules governing the type of installation which is contemplated at the above mentioned location and hereby agree to make the installation in conformance with the Ordinance.

In making this application, I realize I am assuming the responsibility of a licensed contractor for the installation of the work mentioned in the permit application and for putting the equipment in operation. I further agree that I shall neither hire any other person for the purpose of installing any portion of the________________**or related equipment at the above premises, nor sub-contract to any other person, firm or corporation the installation of any portion of the above equipment.**

I agree to notify the Inspection Department within seventy-two (72) hours after the installation is completed and is ready for service so that the Department may make its required inspection. I further agree to keep all parts of the installation exposed until the installation is accepted as being in compliance with Ordinance requirements.

I further agree to correct within two weeks time any violations on the work installed and to provide access to the premises between the hours of 8 a.m. and 5 p.m., Monday through Friday for the necessary inspection or inspections. Failure to correct violations or to provide access will subject the permit to cancellation in which case a licensed contractor must be employed to complete the work.

APPLICATION: ☐ ACCEPTABLE ☐ NOT ACCEPTABLE ________________

White copy to applicant.
Yellow copy retained in office.

Department Representative

If not acceptable, list reason________________________________

Subscribed and sworn to before me this

________**day of**________________, 19____.

____
Owner

____
Present Address

Notary Public,________________County, Michigan.

My commission expires________________.

____
Telephone Number

FIG. 2-3 Homeowners permit for electrical wiring.

testing facility but plays an important part in safety and quality in the electrical industry. Most manufacturers voluntarily submit their products to the Underwriters Laboratories for testing and evaluation. Permission to display the UL label on the manufacturer's products is granted if the product tests satisfactorily. The Underwriters Laboratories does not *approve* products; it only *lists* safe products.

Testing and listing of products are not restricted to electric products alone. Fire prevention products and many industrial products are listed. Items without the UL label are inferior or defective and are not safe to use. Always buy products with the UL seal for safety and quality.

PERMITS AND INSPECTIONS

When adding to a system or making extensive additions such as an added room to be wired, you must obtain a permit from the local building inspection department. You can obtain a *homeowner's permit* from most municipalities. Your work will be inspected and usually approved. If it is not approved, the inspector will advise you on how to correct your work. Figure 2-3 shows a homeowner's wiring permit.

CHAPTER 3

Hand Tools for Wiring

Many of the tools used in electric wiring are common to other trades and used also by the home repair person. Other tools have special designs or perform special functions. If you do many home repairs, you may already have the tools needed for electrical work. The tools you will need can be grouped into two categories: those used to do the "rough-in," the work necessary to install the electric cable, the boxes, the drilling of holes, and the mounting of boxes and panels; and those used to connect the devices, test the system, hang fixtures, and complete the work.

ROUGH-IN TOOLS

Rough-in tools and equipment are more sturdy and heavier and, as the name implies, do rough work.

Carbide Bits

Carbide bits are used for drilling holes in concrete, block, and brick. The smaller sizes are reasonably priced and surely are supe-

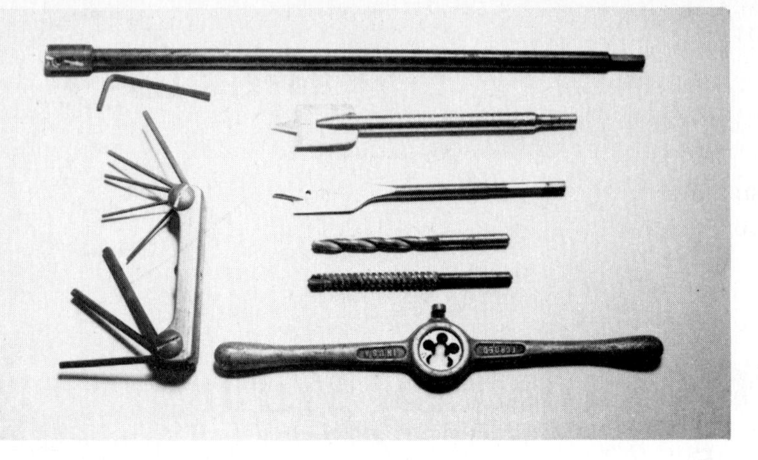

FIG. 3-1 From the top: Spade-bit extension; two spade bits ($\frac{3}{4}$- and 1-inch), high-speed drill bit; carbide bit; die handle and die; and pocketknife-style Allen wrench set.

rior to the star drill and hammer (see Fig. 3-1). For mounting electric boxes on masonry walls, use the $\frac{3}{16}$-inch drill bit and plastic anchor to fit. Fasten with a No. 10 pan head sheet-metal screw. Larger holes through concrete walls may call for the rental of a rotary hammer-type drill motor with a long carbide bit. You may never need to drill large through-holes $\frac{3}{4}$-inch and larger, but this equipment is available at rental stores. Also available is a drill motor that combines straight drilling in wood and metal and hammer drilling in masonry. A ring near the chuck turns to select the desired mode. Small holes for screw anchors avoid the need for the hammer feature. Hammer drills are also available for rent.

Chisels

There are two kinds of chisels—**wood** and **"cold" chisels.** If you are cutting into walls and building framing of wood, you will find wood chisels handy. Cold chisels are tougher and will chip and chop various metals as well as concrete, brick, and block masonry. The **star drill** is similar to the cold chisel except that the cutting edge is in the shape of a cross. To drill holes in masonry, start the drill with care, making sure it is on target; strike the head and at the same time, rotate the drill. *Remember to wear goggles.* This job goes very slowly, as does most hand work. Most holes are now drilled with an electric drill motor and a carbide bit. The cold chisel can be used as an emergency star drill. Use the chisel basically as you would the star drill, except rotate it more often. The hole will look as if you had used a star drill. Many tools can be used for purposes other than those they were designed for. On occasion no tool suited to the job at hand is available, so another tool must be substituted or created. If you are resourceful and handy with tools, you can take care of almost any job.

Conduit and Thinwall Benders

At times conduit or thinwall needs to be bent for the proper fit; in these instances, a tool called a **bender** is right for the job. For short jobs benders may be rented. Fortunately, rigid conduit is not necessary in normal residential work since it requires extra equipment such as threading tools, reamers (see Fig. 3-2), pipe wrenches, and a special bender called a *hickey*. Electric metallic tubing (EMT) is approved by all governing bodies and requires only a hand bender (see Fig. 3-3) and a hacksaw. Benders cost about $16. Working with and bending thinwall tubing are detailed in Fig. 3-4.

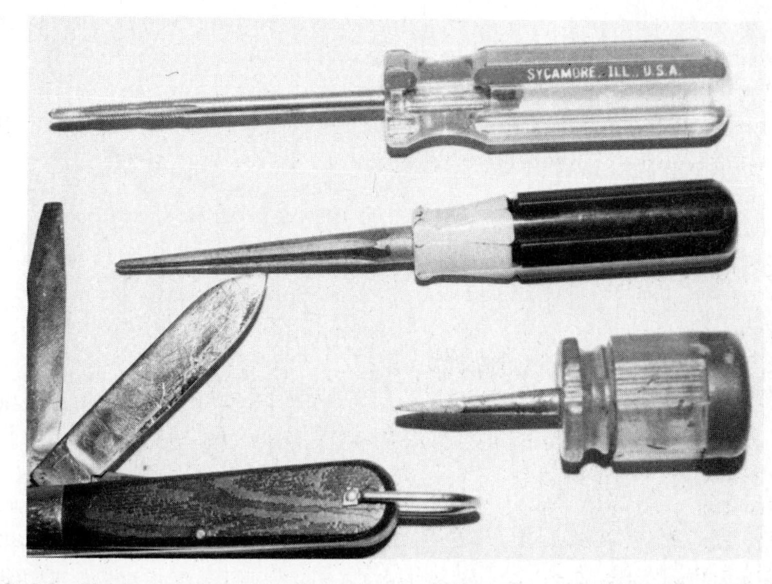

FIG. 3-2 Finish or pocket tools. From the top: Rethreading tool which cleans 6-32, 8-32, and 10-32 threads in boxes; pocket reamer to enlarge holes; stubby screwdriver; and electrician's knife (one knife blade and one locking screwdriver blade).

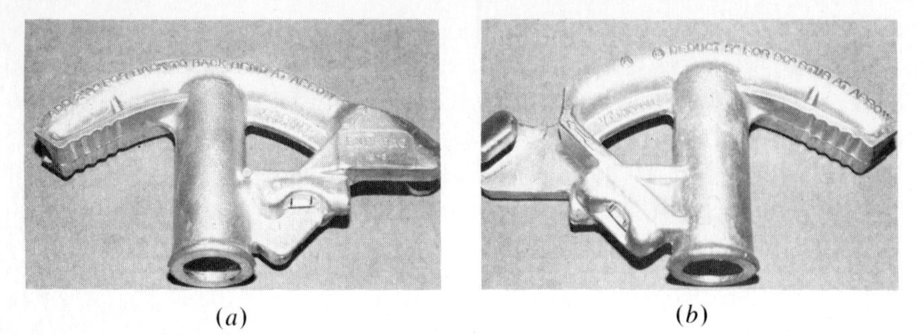

(a) (b)

FIG. 3-3 (*a*) Thinwall bender showing level for 45° bend. (*b*) Reverse side of bender showing level for right angle (90°) bend.

PRELIMINARY PROCEDURE: Read all instructions carefully and completely before attempting to assemble or operate your hand bender. Most malfunctions in new equipment are the result of improper operation and/or assembly.

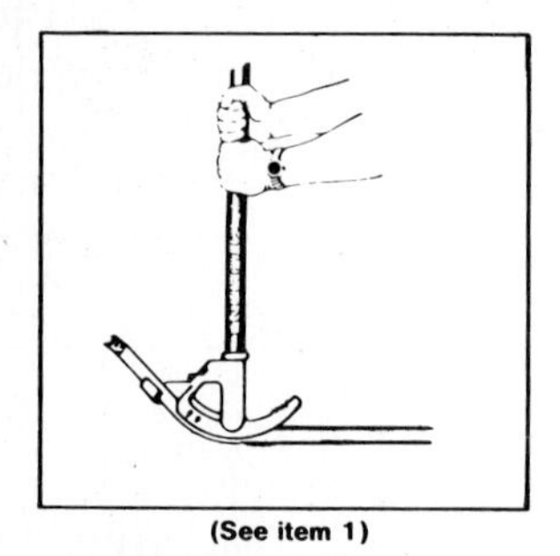

(See item 1)

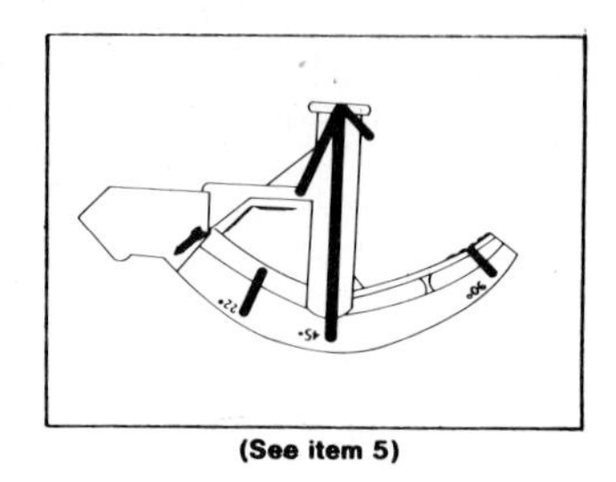

(See item 5)

1. Secure pipe conduit firmly into bender head before attempting to bend.
2. Operator's feet should be well positioned on unobstructed floor or platform before attempting to bend.
3. Keep area around bending clear of any objects which could cause operator to trip or fall while bending.
4. Do not use around exposed electrical conductors.
5. Work on level surface to utilize the built in angle indicators to their fullest.
6. Do not use bender as sledge for removing bridging or other obstructions in path of conduit run.

FIG. 3-4 Instructions for using the bender shown in Fig. 3-3. (*Courtesy of Enerpac.*)

Hammers

Included in the list of hammers are **carpenter's hammers, ball peen hammers** (see Fig. 3-5), **sledges, tack hammers,** and **modelmaker's hammers.** Safety goggles should be worn during pounding or drilling or in any situation where materials could chip, splinter, or drop into the eyes.

> CAUTION: If any material is lodged in the eye, get medical attention immediately!

The carpenter's hammer is most commonly used for electrical work, usually the straight claw style. For pounding on cold chisels and similarly hardened metals, use the ball peen hammer. This hammer is specially made for pounding on hardened metal. Always wear goggles, and use care when striking the chisel with the hammer. Light tapping on a screwdriver handle with electrician's side cutters (pliers), even though neither tool is being used for its intended purpose, is commonly done with no harm to either tool. The same combination is used to tighten locknuts on cable and conduit fittings when they are attached to wall or junction boxes. This is done by tapping the screwdriver head while the point is held against the notch on the locknut edge.

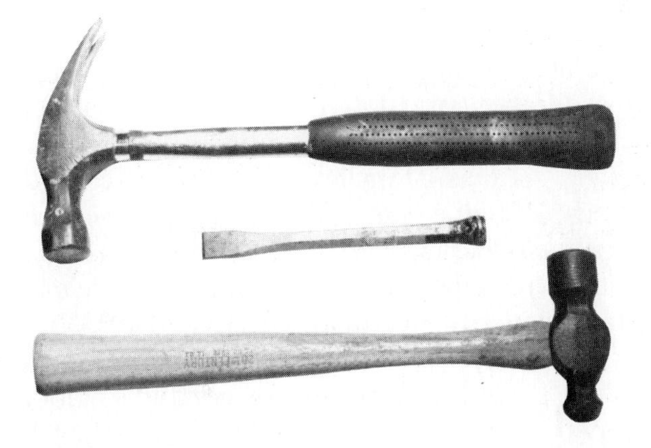

FIG. 3-5 From the top: Carpenter's "ripping" hammer, cold chisel, and ball peen hammer.

Another method of tightening these locknuts is to assemble the locknut hand-tight. Hold the locknut with your free hand while tightening the fitting body with Channellocks. If you end up with the setscrew over to one side, reposition the body and hand-tighten the nut again. All this takes little time to do and seems to be a better way.

Knockout Punches

These **punches,** available from electrical wholesalers, can be used to make additional knockouts in metal boxes only. The two common sizes are $\frac{1}{2}$ and $\frac{3}{4}$ inch. The $\frac{3}{4}$-inch size will enlarge a $\frac{1}{2}$-inch knockout to accept $\frac{3}{4}$-inch connectors. Although they are not absolutely needed, these punches do come in handy at times. For instance, a hole saw will make a new hole but cannot enlarge an existing $\frac{1}{2}$-inch knockout because there is no guide for the pilot drill of the hole saw mandrel. To make a new knockout where none has been, a $\frac{3}{8}$-inch hole must be drilled for the pull bolt of the punch, which pulls the two parts of the punch together. These punches cost $5–$8 each. The Greenlee Company makes them in sets and singly, available at electrical wholesalers. All steel tools are expensive now.

Levels and Plumb Bobs

The carpenter's **level** (2 ft or longer) is familiar to all of us (see Fig. 3-6). Many homeowners have the pocket or "torpedo" level (about 10–12 inches long). Both levels have three vials, each one measuring a different orientation: level, plumb, and 45 degrees. Longer carpenter's levels may have duplicates of some of or all three of the vial settings. Many levels now are made so the vials are easily replaceable. Some plumber's levels can be set "off level" by an adjustment screw to measure different slopes as the slight pitch needed for drainage pipes. Thus when the level reads "level," the pipe on which it is set has the pitch to drain liquids.

The torpedo level is usually sufficient for the electrician's needs. Items to be leveled are cabinets, conduit or tubing, and wall boxes and their devices. Some pocket levels have a groove on the bottom so that they can sit on a horizontal pipe; others may be

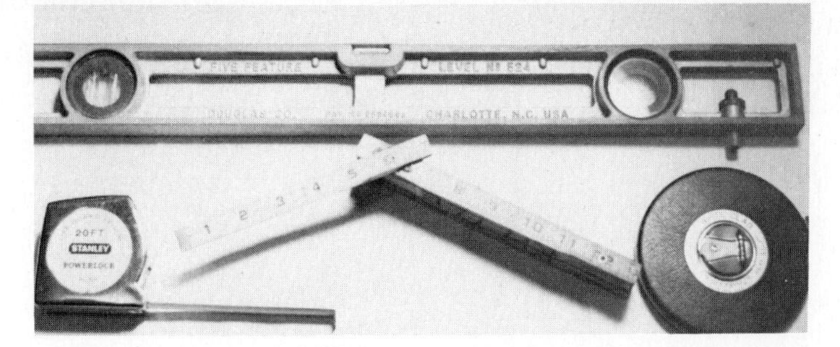

FIG. 3-6 Two-foot aluminum level with an off-level adjustment at the right end; 20-ft tape; wood rule; and 100-ft steel tape (note the windup crank in the center of this tape).

magnetized for use on iron or steel objects. This feature is very handy.

> ***Note:*** Electrical inspectors like to see everything neat and will be more likely to approve the work if it looks well balanced. Everything will be inspected anyway, but the inspector will recall your neatness. Remember that the inspector is always willing to advise you and answer questions about your work. Do not be afraid to ask questions if you are unsure about anything.

The **plumb bob** is similar to the mouse mentioned on p. 49 (see Fig. 3-7). It is a steel or brass cone-shaped weight suspended from a chalk line. A vertical line can be located on a wall by suspending the plumb bob by its line from a point on the upper wall; allow the bob to stop swinging and then mark the location of bob's point.

Measuring Tools

Common measuring tools are the **steel tape** and **wood folding rule** (see Fig. 3-6). Steel tapes come in various widths and lengths. Some tapes are claimed to extend a certain distance, but whether they actually extend this far without bending is sometimes questionable. Wood rules come in lengths up to 10 ft. Three sections can be bent at right angles and used to measure horizontally on the ceiling. This procedure will give only an approximate measurement

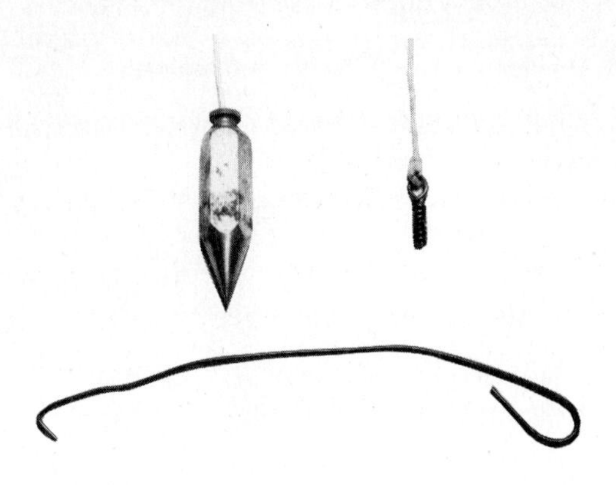

FIG. 3-7 Plumb bob, "mouse," and fish tool.

but will save climbing a ladder. When working on the floor, sometimes you can use your wood rule as a hook to get out-of-reach objects. Just bend the first section at a right angle, and extend the rest of the rule like an arm; this will save you from getting up and down again. This is better known as the lazy man's method.

Wood rules are available in *outside-reading* and *inside-reading* styles. With outside reading, the numbers starting with 1 are visible when the rule is folded up. With inside reading, the numbers ending with 71 are visible when the rule is folded up. The inside-reading rule will lie flat when it is opened partially to, say, 30 inches for a 20-inch measurement.

Steel tapes also come in 50- and 100-ft lengths. In these tapes, a crank on one side of the case rewinds the tape into the case. Try to avoid dragging the tape through sand and over rocks during rewinding. Even though the entrance slot of the case has rollers, hold the opening so that the tape does not have to ride either roller. Rotate the case forward a bit to overcome the drag on the tape.

Power Tools

The **power drill,** also known as the *drill motor,* has revolutionized the construction industry. The increase in work output has been tremendous. The best all-around drill motor is the $\frac{3}{8}$-inch reversing, variable-speed type. This drill is made usually in the *double-insulated* style, having only two prongs on the attachment plug. These are listed by the Underwriters' Laboratories (UL) and are satisfactory.

> ***Note:*** These double-insulated tools must be repaired by an authorized service center for that make of drill as assembly is technical and must be done with proper testing equipment.

The $\frac{1}{4}$-inch drill motor is too light, and the $\frac{1}{2}$-inch motor may be too heavy for your needs. Buy well-known brands such as Black & Decker, Skil, Rockwell, Craftsman, and Montgomery Ward. Notice the old Yankee push drill in Fig. 3-8.

Saws

Saws are classed as either **wood saws** or **metal-cutting saws,** such as the hacksaw. There are also masonry saws with carbide teeth, but they are not within the scope of this book. Wood saws include the carpenter's **rip** and **crosscut saws. Keyhole** and **drywall saws** are similar to the carpenter's saws. They are used for rough work such as cutting plaster and lath and drywall (see Fig. 3-9). A master carpenter would never use wood saws for these purposes because they would be dulled and otherwise damaged. The drywall saw has a screwdriver-type handle, and the keyhole saw has a handle similar to that of a wood saw.

There are saws that cut round holes in drywall and soft metals such as aluminum. These saws, called **hole saws** (see Fig. 3-10), are used in an electric drill motor and range in size $\frac{1}{2}$–4 inches in diameter. In addition to the saws mentioned for cutting soft materials, the better-quality hole saw is known as the *high-speed* saw, just as twist drills are designated. These saws will cut holes in

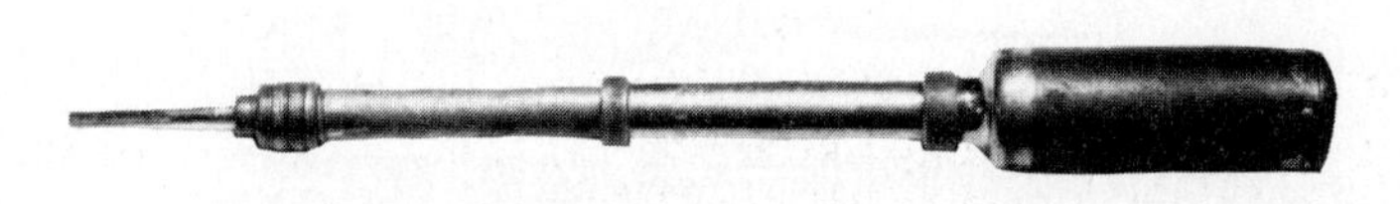

FIG. 3-8 Yankee push drill. This is over 40 years old and still works well.

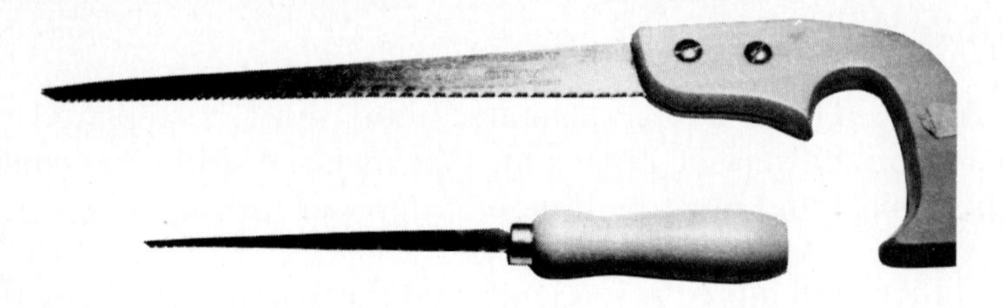

FIG. 3-9 Keyhole saw and drywall (wallboard) saw.

FIG. 3-10 From the left: $\frac{5}{8}$- and 1-inch-hole saws; $\frac{1}{2}$-inch nominal knockout punch (actual diameter is $\frac{7}{8}$ inch).

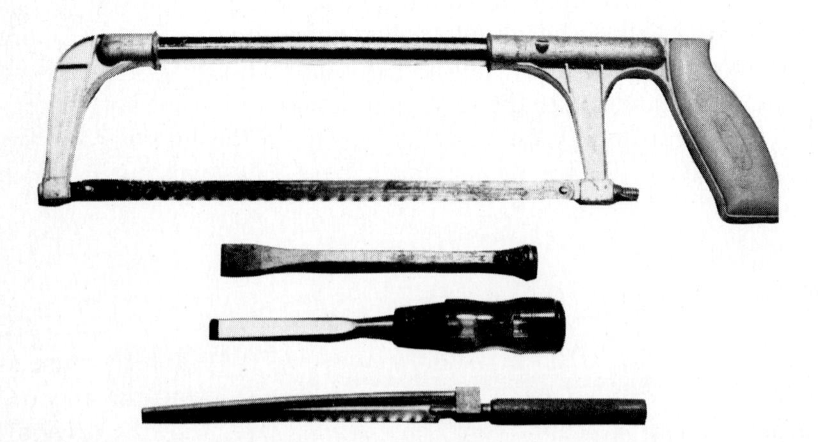

FIG. 3-11 From the top: Standard hacksaw frame with blade which will accept either 10- or 12-inch blade lengths; cold chisel and a wood chisel (for comparison); and pocket hacksaw.

metal of all kinds. Stainless steel is extremely hard but can be cut and drilled with the high-speed twist drills as well as with the high-speed hole saws. To accept the wide range of saw diameters, two mandrels (the part with the pilot drill and the other end to be chucked in the drill motor) are used; one ranges from $\frac{5}{8}$- to $1\frac{1}{2}$-inch diameter, and another ranges from $1\frac{3}{4}$- to 4-inch diameter.

Hacksaws, designed for sawing metals of all kinds, are found in most home workshops—both the **frame** type (see Fig. 3-11) and the **keyhole** type (similar to the wood keyhole saw). Another keyhole type will take broken frame-type hacksaw blades. The best blades are also called *high-speed* and have flexible backs and hardened teeth that last a long time. They cost about $1 each but are worth the price. The cheap blades will break at the least strain.

Spade Wood Bit

The **spade wood bit** for use with a drill motor is good for boring rough holes in studs and other framing members. The $\frac{5}{8}$- or $\frac{3}{4}$-inch size is commonly used. There are 18-inch extensions for use with the spade bits (see Fig. 3-1). The bit fits in the socket on one end and is secured with an Allen setscrew. The other end of the extension fits in the drill chuck. Both the drill bit and the extension have flats which must be lined up with the Allen setscrew or the drill chuck jaws for tightness. Spade bits cost $1.50–$3, depending on the size. The extension costs $5–$8. Check the surplus stores in your area for useful tools at lower prices. My extension cost $1.50; it is not chrome-plated, but is rustproofed. Be sure to check for good quality. Again, it is best to buy the better-known U.S.-made tools. Some surplus tool stores do have good-quality government tools at low prices.

Struck or Hammered Tools

Cold chisels, star drills, punches, and steel-handled wood chisels are known as *struck,* or *hammered,* tools. The working end may be sharpened, and the hammered end may be dressed to the original shape. Bevel the end that is hit to prevent mushrooming, because small pieces of the mushroom shape may break off and injure you.

CAUTION: Do not use a carpenter's hammer to strike these tools. The tool head may be small and chip the hammer head. Use a ball peen hammer or small slege or mall. *Always wear goggles* or other eye protection when you are using any tools! This is a new and important safety consideration. Many new tools have this warning stamped into the tool itself. Although goggles may be uncomfortable and inconvenient, so is seeing with only one eye or being blind. Always seek immediate aid for any injury, especially eye injuries.

Threading Tools

At times small bolts or other bolt-type cylinders have defective threads and no replacement is available (as on a Sunday evening when the stores are closed). At such a time it would be nice to be able to rethread (chase) the damaged threads yourself and not have your work held up. Common threads used in electric work are 6-32, 8-32, 10-32, 10-24, and $\frac{1}{4}$-20. Available **dies** to cut these threads are 1 inch in diameter (see Fig. 3-12). A handle is sold to hold these dies for use. In this case, a bench vise or Vise-Grip pliers are used

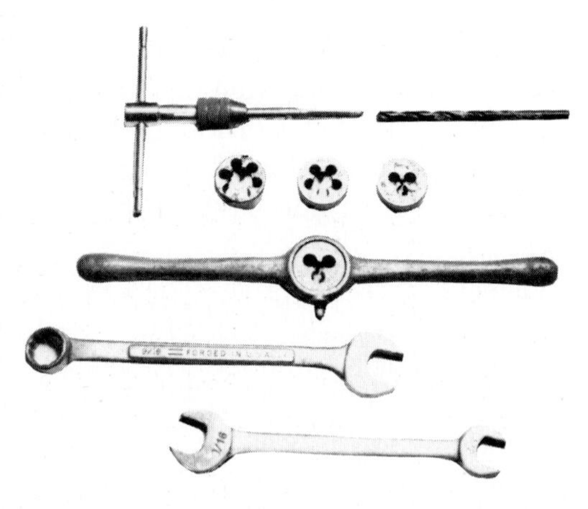

FIG. 3-12 From the top: Tap wrench holding a tap, with matching high-speed drill bit (each tap takes a specific drill size to match the tap size); dies and a handle with a die in place; "box-open" wrench, $\frac{9}{16}$ inch on both ends (for nuts measuring this distance across their flats); and "open-end–$\frac{9}{16}$-half."

to hold the bolt. If you're just reconditioning the threads, no cutting oil is needed to help the process. Cutting oil *is* needed to cut new threads. Any available oil such as motor oil or even cooking oil or lard can be used.

Matching **taps** (for threading internally) in the sizes mentioned above come in sets or are sold individually, as are the dies. Sometimes wall-box mounting holes or other holes are defective or have not been tapped at the factory. The box may already be in place, in concrete perhaps. To remove and replace it would be time-consuming, but the correct tap would remedy this in a minute. Wall boxes have 6-32 tapped holes, ceiling boxes have 8-32 tapped holes, and grounding screw holes are 10-32 tapped holes. Brands recommended are Hansen, Ace, Greenfield, and Craftsman. All these are high-speed steel and are top-quality. A new tool shaped like a screwdriver has taps 6-32, 8-32, and 10-32, each behind the other smaller one, all on one shaft. This is nice to have. Both taps and dies cost about $1 each, and the combination tool costs about $3.

Tool Pouch

A small **tool pouch** can protect your hip pocket from wear. I would suggest a pouch holding four or five tools only. Good pouches of leather cost about $10.

Vises

The need for a **vise** in electric work is small because most work is done on the job site instead of at the workbench. Occasionally metal junction boxes need additional holes for screws or connector fittings. The home workshop vise sells for $20–$30. Sometimes it is possible to cut costs by purchasing a used vise at a garage sale. Be sure to check the operation of the vise before buying it. Many workshop vises have removable pipe jaws that might be useful.

Wrenches

All tool categories have dozens of styles, many with specialized uses. Some tools are designed to the specifications of certain companies. The telephone companies have many such tools. Wrenches are no exception.

Wrenches that you might need for electric work include the following:

- An adjustable **Crescent-type wrench** (see Fig. 3-13).
- Some small ignition-type **open-end wrenches** (handy for appliance repair). A Crescent wrench can be adjusted to fit various size nuts whereas an open-end wrench is fixed in size.
- A ($\frac{1}{4}$-inch) drive **socket set** having a screwdriver-type handle to accept various sizes of sockets (also handy for appliance repair).
- A small set of **box wrenches** for hex nuts. A box wrench is in the form of a ring which fits *over* the hex nut, which is different from the open-end wrench that slides onto the nut from the side.

Vise-Grip pliers can be classed as wrenches because they clamp onto nuts and other parts and act as a vise. Buy the Vise-Grip brand. Although these and other quality tools may look the same as other tools, their advantages will become evident as you work with them. Even their costs may be similar, but look for the recommended brands—they're much superior.

You might have use for one or two small **pipe wrenches.** The 10-, 12-, or 14-inch sizes are suitable. The Ridgid brand is best, but it is expensive. You may prefer to buy the cheaper (but still sturdy) brands to save money. Pipe wrenches also may be rented, if necessary, for a one-time use.

Allen Wrenches

Allen wrenches are necessary when Allen-head bolts or setscrews are worked on. Actually nothing else will work on a tightly secured Allen-head bolt or setscrew. Wrench sets are now made in the pocketknife style, with each end having a number of different size wrenches arranged as the blades of a pocketknife. These sets come in different size ranges depending on your needs. A mill supply house or many hardware stores have individual wrenches in various lengths. Only occasionally will you need an Allen wrench for electric work [some large terminal lugs will have Allen setscrews (see Fig. 3-1)].

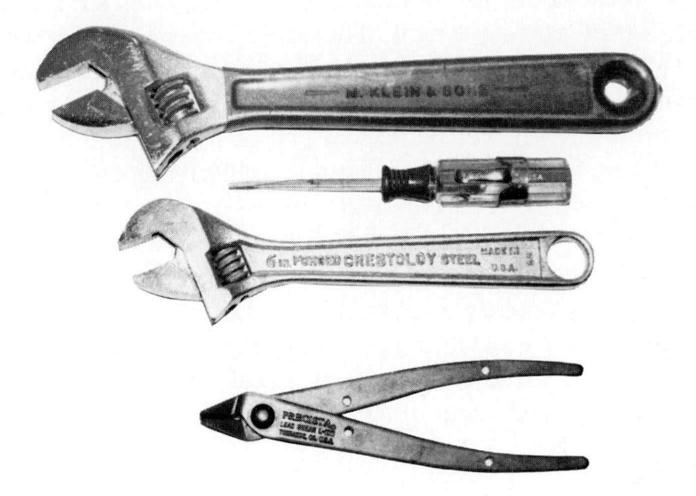

FIG. 3-13 From the top: Crescent-type wrench, 8-inch; pocket screwdriver; genuine Crescent wrench, 6-inch; and small diagonal cutters.

CAUTION: Pay attention to *where* you are using your metal tools. *Never* use an Allen wrench on hot (live) terminals. Some panel covers may have Allen screws holding them in place.

TOOLS FOR FINISH WIRING

Tools in this category are used to install switches and receptacles and to hang fixtures. Generally these are pocket or tool-pouch tools.

Mouse

The **mouse** (see Fig. 3-7) is a handmade tool which is wire solder formed into a compact weight to be dropped, on a chalk line, down inside a hollow wall to be retrieved through an opening in the wall surface by means of a wire hook. A 3-inch length of wire solder is bent back on itself, leaving a loop; then the rest is wound back on itself to form a compact, heavy object.

The **pocketknife** is a handy tool for stripping wire insulation and for other cutting needs. The knife must be used with care because it slips easily and cuts not only the material but also fingers and hands. An *electrician's knife* is available which has one large blade and a screwdriver blade that can be locked in the open position (see Fig. 3-2). This knife is not absolutely necessary since the pocketknife and screwdriver will suffice.

A section of coat hanger wire bent with a loop on one end for a handle and a short right-angle hook on the other end is used to fish a mouse out of a hollow wall opening. This makeshift tool is called a *fishhook*. Many special tools such as the two described above can be fashioned to suit the need. Even small open-end wrenches can be made from sheet metal for a special use.

Pliers

There are as many styles and types of pliers as there are manufacturers. Most well-known brands are of high quality. Good tools are made by Sears, Wards, True-Value, Crescent, Channellock, and Klein Tools. Klein Tools makes a complete line of tools especially

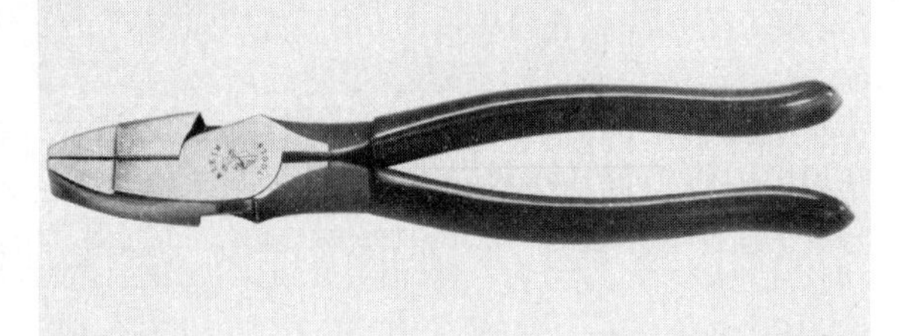

FIG. 3-14 Electrician's side-cutting pliers. (*Courtesy of Klein Tools.*)

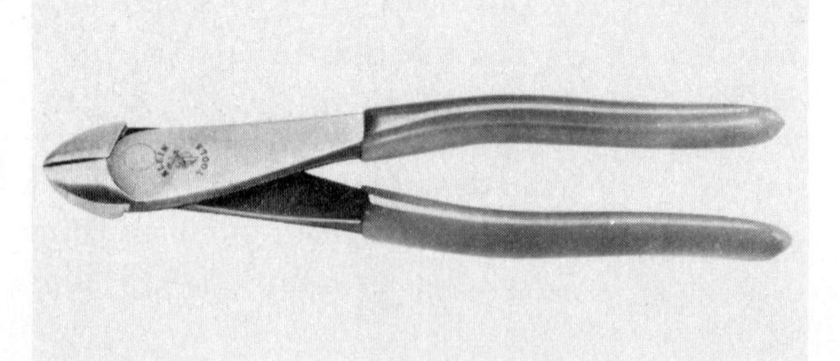

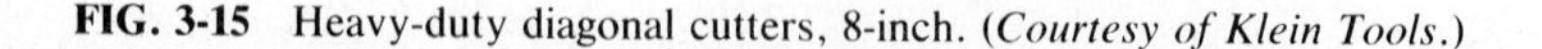

FIG. 3-15 Heavy-duty diagonal cutters, 8-inch. (*Courtesy of Klein Tools.*)

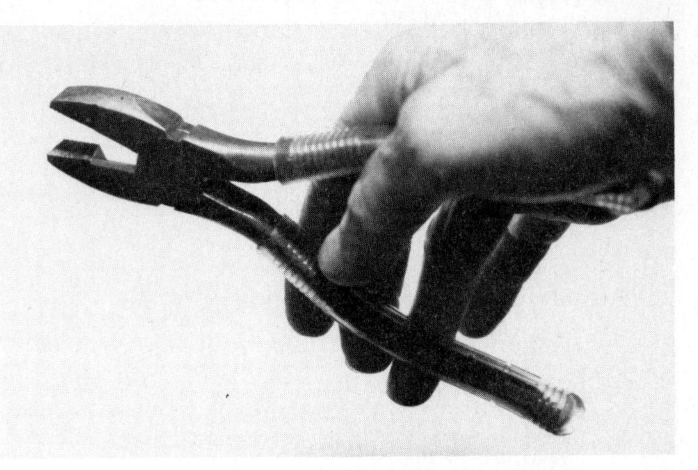

FIG. 3-16 The correct way to hold side cutters.

for the electrician. These are sold by electrical wholesalers.

Two of the pliers most often used by electricians are the **lineman's side-cutting pliers** (Fig. 3-14) and the **diagonal pliers** (Fig. 3-15). The proper way to hold side cutters is shown in Fig. 3-16. Use the thumb and base of the forefinger to hold onto the handle; allow the other side to drop open by gravity. If pliers are kept well lubricated, there is no need to pry them open with your other fingers.

Diagonals, as opposed to side cutters, are quite handy in that they will cut all the way to the tip. Also popular are **long-nose** (needle-nose) **pliers** (Fig. 3-17), which are used for forming loops on wire ends and reaching into narrow places. Specialized pliers that crimp terminals on wire ends, cut machine screws, and strip insulation from wires are also available.

For **arc-joint pliers** (Fig. 3-18) a favorite is the Channellock brand. This company makes a full line of tools also. All well-known manufacturers' tools look similar, but certain tools seem to work better. Small differences in engineering design cause the tool to operate better than that of the competitor. Channellock arc-joint pliers are an example of superior top design. Although Channellock pliers are the best designed you may not be able to tell just by looking at them, but you'll probably find that they work better than all other brands.

Many pliers and other tools such as snips have colored plastic grips on the handles. This dipped coating must *not* be thought of as

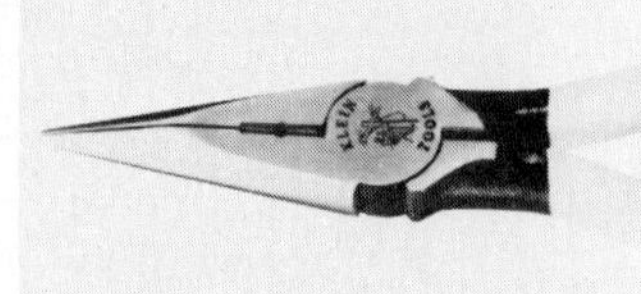

FIG. 3-17 Long-nose pliers, 8-inch. (*Courtesy of Klein Tools.*)

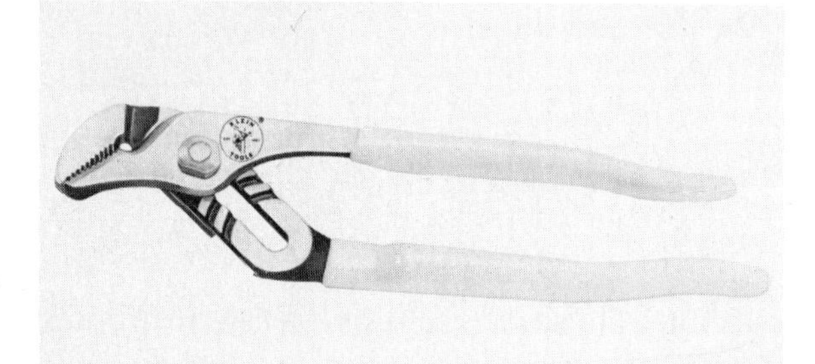

FIG. 3-18 Arc-joint pliers, 10-inch. (*Courtesy of Klein Tools.*)

insulation from electricity. Klein Tools states, "Plastic dipped handles are *not* intended for protection against electrical shock."

Screwdrivers

The **screwdriver** is one of the most common tools in use. Although a special screw may require a special screwdriver, most screws on electric equipment require either the **slotted screwdriver** or the **Phillips-head screwdriver** (see Figs. 3-19, 3-20, and 3-21). However, with any type of screw, the size of the screwdriver blade *must* fit the screw slot. This point is especially important to remember when you work with slotted screws since the wrong screwdriver size can sometimes be forced to fit the screws. Trying to make do with the wrong size tool can damage the screw, the screwdriver, or both. Usually three sizes of screwdrivers will fit most screws on electric equipment. One choice would be $\frac{3}{16} \times$ 4-inch blade, $\frac{1}{4} \times$ 4- or 6-inch blade, and $\frac{3}{8} \times$ 6-inch blade. Or No. 1 and No. 2 Phillips-head screwdrivers might be needed. All handles should be plastic with a rubber grip over the plastic; if you have trouble finding them, try the wholesaler.

One of the most useful screwdrivers is the **screw-holding screw-driver.** There are many types, but the best is the Quick-Wedge brand. This type holds the screw by expanding the thickness of the blade that goes into the screw slot, thus holding it firmly. Electricians use this type to work on hot connections because the blade and handle are insulated. The screw can be inserted, snugged up,

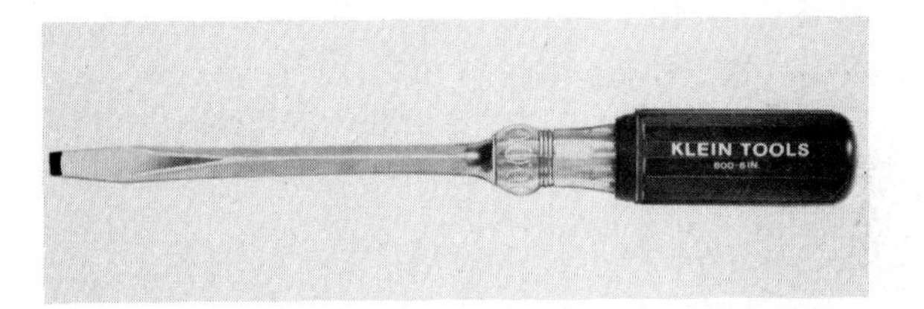

FIG. 3-19 Square-shank screwdriver, 6-inch. (*Courtesy of Klein Tools.*)

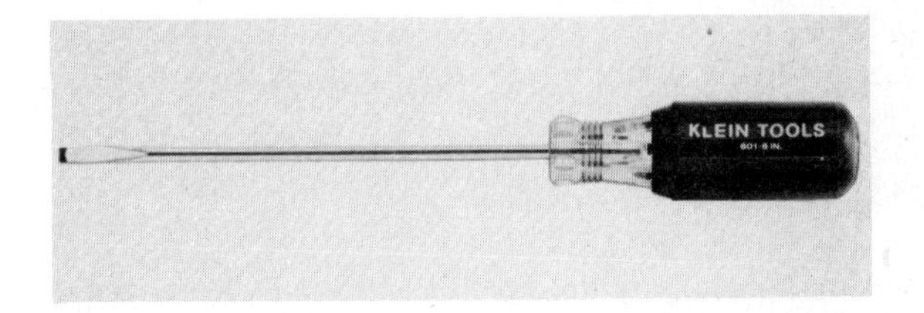

FIG. 3-20 Cabinet tip screwdriver, 8-inch. (*Courtesy of Klein Tools.*)

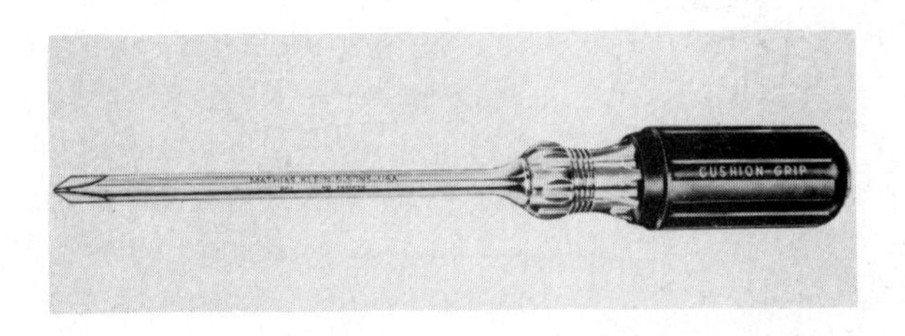

FIG. 3-21 Phillips-head screwdriver, 8-inch. (*Courtesy of Klein Tools.*)

and released with one hand which touches only the insulated part of the screwdriver.

Do not use this screwdriver to either loosen or finally tighten screws. It is designed for turning loosened screws only. Putting too much torque on the split blade will damage the screwdriver.

A **pocket screwdriver** is handy for when you need to work with small screws. The **stubby-style screwdriver** is also useful. This is not to be confused with the screwdriver having the ball-shaped handle. This tool is not suited to our purposes.

Always use insulated-handle screwdrivers, even when you are not working on hot wires. Many screwdrivers have a rubber sleeve over the plastic handle to cushion the hand. Do not buy a plastic-handle screwdriver with a steel mushroom-shaped top on the handle. This steel top runs through the handle and is part of the blade. If you were to tighten a hot screw with it, you would get a shock.

> CAUTION: Be careful when you are working *near* hot wires, such as in a distribution panel. If you have turned off the main breaker and are tightening all terminal screws (including the two hot screws on the incoming side of the main breakers or fuse blocks), practice holding the screwdriver by the plastic *only*. Make it a habit *never* to touch the metal shank.

Finally, spare no expense and buy high-quality tools, which are a good long-term investment.

Slitter for Romex

This tool, used to slit the outer covering of Romex to expose the insulated wires inside, is very handy. At under $2 it saves time and minimizes the chance of cutting the insulation on the wires. It is hooked over the cable and squeezed together. A pointed blade inside the tool can be used to slit the cable covering. Sometimes the blade must be sharpened to cut properly, even when new.

TEST EQUIPMENT

Electric test equipment comes in various price ranges. Your choice depends on your particular needs, starting with **pocket voltage testers** that cost under $2. Watch what you buy because there are also *low*-voltage testers used for automobiles. These are rated 0–50 V and will burn out if they are used on household voltages. When you buy any piece of equipment, always read the information first. It may save you the time and trouble of returning the item for refund or replacement.

A second tester, called a **plug-in lamp socket adapter,** consists of a lamp socket with plug-in prongs at the opposite end. Screw in a bulb and then push the assembly gently into a receptacle. Make only slight contact; do not push all the way in. This avoids pressure on the bulb, which might break if it were pushed hard. Momentary contact shows whether the circuit is hot. This item costs under $1 (see Fig. 3-22).

Larger professional testers called **voltage testers** cost about $20 and provide longer leads. An indicator pointer moves down a vertical slot in the tester front to indicate the voltage being tested. This tester, a pocket style, is made to take the constant usage an electrician will give it.

> ***Note:*** The voltage shown is not exact, as is shown on a true voltmeter; it only indicates the voltages as 120, 240, 480, and 600 V.

The tester shown in Fig. 3-23 is Square D; other makes are Ideal and Sears (Craftsman).

Most **volt-ohmmeters** (VOMs) will register ac volts but not ac amperes. The VOM is used mainly in electronic work; it indicates ac volts accurately. This tester can be used in another way that is explained later in this section. The simple small VOM sells for $15. Others can run in the hundreds. The VOM shown in Fig. 3-24 is by EICO, Inc.

The four testers mentioned will test for 120 V. The professional voltage tester and the VOM will test to 600 V, which is all you will need. If you work with electronic equipment, you will have a VOM already.

Another type of tester is the **continuity tester.** This tester tests only the completeness of electric circuits. This tester must be used *only with the power off!* All continuity testers are battery-operated. Some are in flashlight form. One pocket type has a test lead on one end. The lead is touched to one end of a wire, and the flashlight head is touched to the other end of the wire. If the wire is good (continuous), the flashlight will light.

Another flashlight tester made by Brite-Star Company is a

FIG. 3-22 Receptacle voltage tester. Prongs plug into a receptacle to show whether power is present.

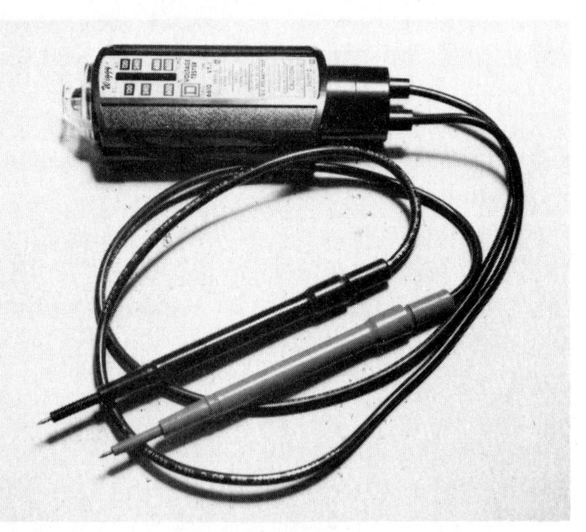

FIG. 3-23 A pocket voltage tester, used daily by electricians.

standard full-size flashlight (see Fig. 3-25). The rear cap has a radio jack, and the two test leads plug into this jack by means of a radio "plug." The flashlight is turned on by its switch but will not light until the two test leads are touched together *or* are touched, one lead to each end of a wire or circuit that is good. When the testing is finished, the plug is removed from the jack in the flashlight. The

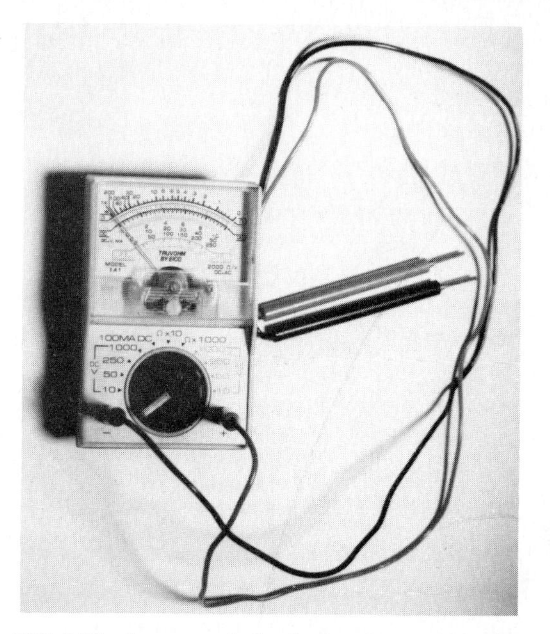

FIG. 3-24 An economical volt-ohmmeter sells for $15.

flashlight is now on and must be turned off by using its own switch. It is now a regular flashlight. Given these two uses, this is a good buy for about $10, for it is industrial quality.

The VOM also tests for continuity by using the ohms scale; an infinity reading indicates continuity. There is a disadvantage in using a meter-type continuity tester: You have to watch the meter needle at the same time as you are holding the test prods on terminals or wire ends. With the flashlight type, you can see the light flash out of the corner of your eye while you use the test prods. This is your choice.

Electricians use a homemade tester consisting of a 6-V lantern battery, a doorbell buzzer, and long test leads. This is a heavier-duty tester because its voltage enables it to test longer circuits on new construction. Usually power has never been applied to these circuits, and this tester is used to check for proper wiring connections. The buzzer sounds to indicate continuity.

> **CAUTION:** Never use these continuity testers on live (hot) circuits. The circuit *must* be dead.

Another tester developed expressly for electricians and persons servicing electric equipment such as motors is a hand-held

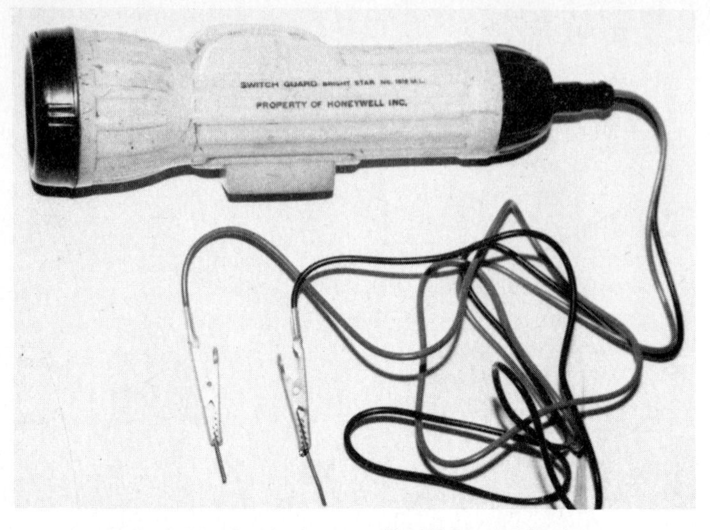

FIG. 3-25 Continuity tester/flashlight combination.

meter having movable jaws at the top. This is called a **clamp-on volt-ohmmeter.** By pressing a lever on the side, these jaws will open. When the jaws are closed, they can encircle a single wire leading to a motor. No actual connection is made as current is induced in the jaws, thus giving a reading on the meter. Voltage is checked with test leads, as with other testers. This meter is convenient for service work on heating and air conditioning. The tester shown in Fig. 3-26 is made by Amprobe. A similar tester is sold by the Grainger Company that is very similar to the one sold by Amprobe except that it is yellow and carries the name "Dayton."

Terminal Crimping Tool

This tool is known as a **combination crimping tool** (see Fig. 3-27), and it can crimp wire terminals on stranded wire, cut wire and small bolts, and strip insulation from wire. Buy U.S.-made tools—all have excellent guarantees, many have lifetime breakage guarantees. All such brands are about equal, but be sure the tool has the capabilities listed above. (Amp, ITT Holub, and Klein Tools are recommended.) Most of the small wire stripping tools on the market strip insulation by using V-shaped notches instead of semicir-

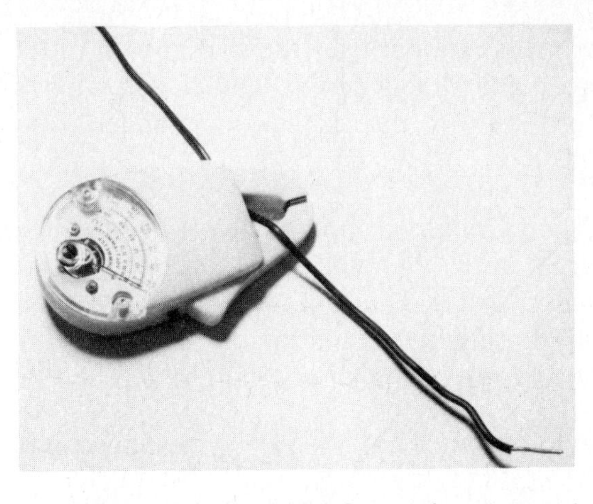

FIG. 3-26 A clamp-on VOM. Current draw (amperes) can be read by clamping around a single wire, as shown.

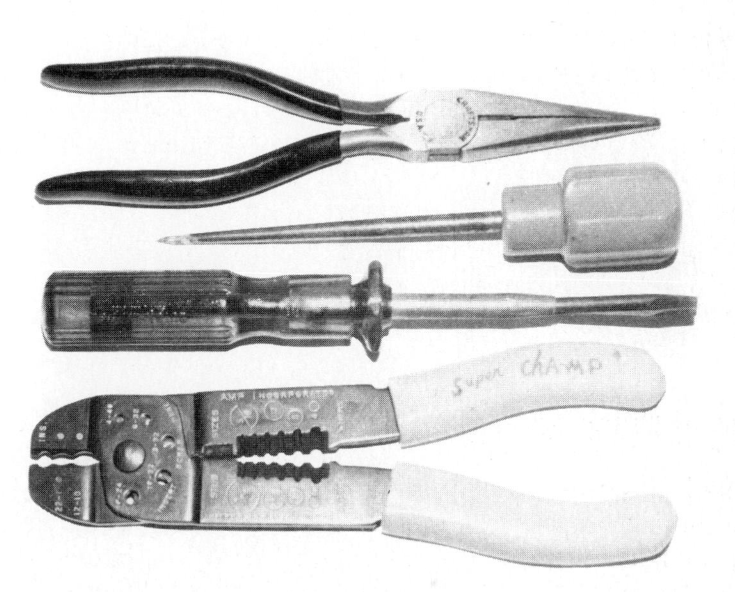

FIG. 3-27 From the top: Long-nose pliers; scratch-all; screw-holding screwdriver, wedge type; and combination crimping tool. Note the crimping tool's four features from the left: terminal crimper, bolt cutter, wire cutter, and wire stripper. All these tools are finish, or pocket, tools used for installing switches, receptacles, and so on.

cular notches. This stripper uses the same notch for all wire sizes by employing an adjustment screw. This V notch tends to nick the wire or cut off some of the strands of stranded wire and is not recommended.

> ***Note:*** When you are stripping wire of unknown size, start by using a larger notch than you think is needed. Close the tool on the insulation, rotate the tool about the wire somewhat, then pull slightly to slide the insulation off. If the insulation does not slide off easily, try a smaller notch until you find the right size.

The semicircular notches are good for stranded wire because they do not cut the strands as readily.

CHAPTER 4

Wiring Materials and Devices

Wiring materials include wire, cable (two or more insulated wires enclosed in an outer sheath), conduit and tubing, supporting and anchoring devices, and enclosures, such as panels and wall and ceiling boxes.

WIRE

Wire sizes range from lamp cord to wire as thick as a large finger. Lamp cord, No. 18 AWG, consists of many fine strands for flexibility. The most common type is called *rip lamp cord* because the two parallel insulated wires are easily ripped apart for making connections at their ends. This cord is used strictly for lamps, clocks, shavers, and light-duty small appliances requiring low wattage. A heavier cord with thicker insulation and larger wire size, No. 16, is used on toasters, TVs, pressing irons, and high-wattage hair dryers. There is a cord made especially for vacuum cleaners. The largest cord is needed for electric ranges and clothes dryers. This may be No. 8 AWG.

Some stores have replacement cords with the attachment plug molded on. If you just need to replace the plug, buy the "dead front" type now required by the NEC. This plug opens up to be wired, because the end with the prongs must not have any exposed terminals or wires. (Bare wire strands might contact a metal wall plate.) Certain appliances have *polarized plugs,* meaning that one prong is wider than the other and the plug can be inserted in the standard receptacle in only one way. Certain electronic equipment will not work properly unless the plug is polarized.

Construction wire used by homeowners usually will be cable. Nonmetallic sheathed cable is commonly called **Romex.** Certain municipalities will allow only **BX** cable, which is used just as Romex, but the outer covering is spiral wrapped steel and is less subject to damage than Romex. The correct name for BX is Armored Bushed Cable (ABC). Both cables come in various sizes and have either two or three insulated wires inside. Sizes No. 10, 12, and 14 are common in dwellings. Three-wire cables are used mostly between three-way switches since three wires are needed between the switches.

Both cable types now have a bare ground wire built into them, copper for Romex and aluminum for BX. The NEC requires this. Romex can be cut with electrician's side cutters or diagonal cutters whereas BX requires a hacksaw or a special, expensive tool (see Fig. 4-1). Then the wires are partially stripped and are ready for connection (see Fig. 4-2). As with any construction work, every installation must be secured and supported in a workmanlike manner. Both types of cables must be stapled every 4½ ft and within 12 inches of an entrance into a box or cabinet. An exception to this rule occurs when the cable is fished down inside a finished wall where staples cannot be nailed. Clamping fittings either built in or inserted into an open knockout grip the cable. Clamps are different for each type of cable. BX cable must use a special bushing inserted between the insulated wires and the cable armor to keep the sharp edge of the armor from cutting the insulation. This is required by the NEC.

Rigid conduit must be cut and threaded like water pipe, and you will not be using it in home wiring. Electric metallic tubing (EMT), commonly called *thinwall,* is bent by using a special thinwall bender. This tubing is easy to bend, but it takes some practice (see Fig. 3-13). Instructions are also furnished with the

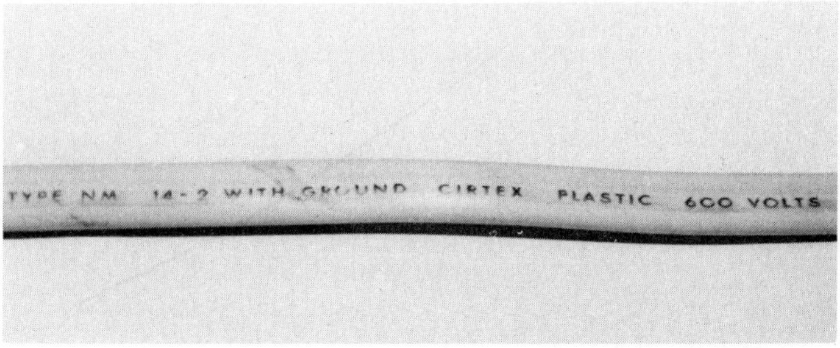

FIG. 4-1 Section of nonmetallic sheathed cable showing required information: 14-2 with ground, 600 V.

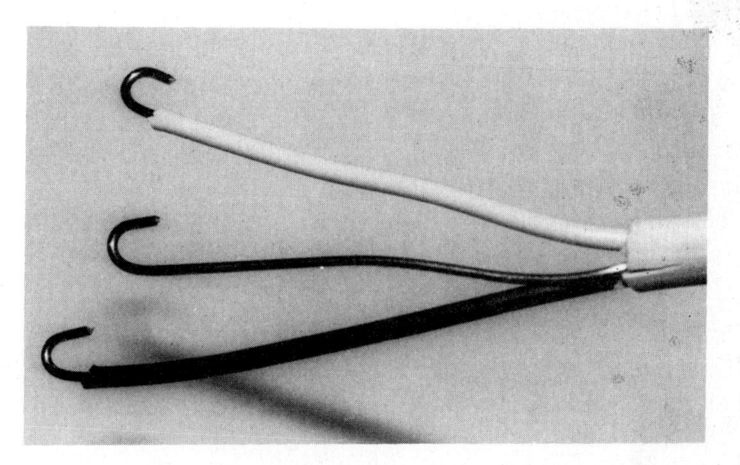

FIG. 4-2 Cable in Fig. 4-1 with ends bared ready for connection to switch or receptacle. The bare ground wire is in the center, and the grounded (white) circuit wire is at the top. (The white wire is sometimes mistakenly called the neutral.) The hot wire is shown at the bottom.

bender. Thinwall uses its own special fittings to connect it to boxes and panels. One new fitting is offset so that the tubing run lies against the wall or ceiling and the fitting end is raised and fits the height of the knockout opening. A connector is needed anyway, and this offset fitting costs little more.

Wire known as *building wire* is used with thinwall tubing; it is stranded for flexibility while it is being pulled through the tubing and around bends in the tubing. Many stores sell this wire by the foot. You will need both black and white wire. You may find only solid wire, which will be alright for your needs. Be sure to buy

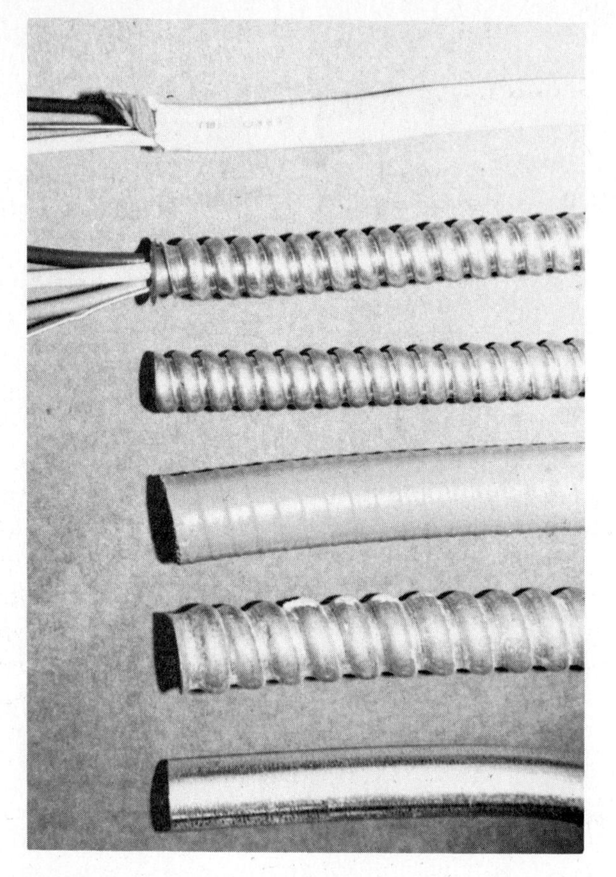

FIG. 4-3 Types of wiring materials. From the top: Nonmetallic sheathed cable, 14-2 with ground; BX, 14-3 with ground, wires are black, white, red, and ground strip; greenfield flexible conduit, $\frac{3}{8}$-inch; Sealtite waterproof conduit; greenfield, $\frac{1}{2}$-inch; and EMT thinwall. (The last four conduits are empty; once they're installed, wires are pulled through them.)

enough to make the connections at each end of the run.

Other conduits, actually called *raceways,* are shown in Fig. 4-3. Not shown is the rigid nonmetallic conduit; this is bent by using heat and special dies. A conduit which looks like BX cable, but without any wires, is called *greenfield*. The common size, $\frac{1}{2}$-inch nominal, is about $\frac{7}{8}$ inch in diameter; it uses the same size BX-type fittings. (*Nominal* size designations are taken from an approximate measurement of the inside diameter of a pipe, conduit, or tubing.) A smaller size is used in special cases where flexibility is needed, such as a motor on a furnace blower. Lengths up to 6 ft are allowed for such purposes. This size is designated as $\frac{3}{8}$-inch nominal. These

FIG. 4-4 Closeup of Sealtite waterproof flexible conduit.

two sizes of greenfield are shown in Fig. 4-3. Also shown is Sealtite flexible watertight conduit, $\frac{1}{2}$-inch nominal size. Sealtite is similar to greenfield because it is strengthened inside by spiral metal covered with a plastic coating (see Fig. 4-4). Special compression fittings attach to the ends of the conduit. Then these fittings are screwed into cast-aluminum boxes to make a watertight assembly.

The same building wire used in the thinwall and rigid conduit is used in these flexible conduits. It is best to use stranded wire. Make a return loop on the wire ends to allow the end to ride over the bumpy inside of the conduits. If there are many bends, a fish tape may be needed to pull the wires through.

SERVICE ENTRANCE CABLES

Service entrance (SE) cable is the heavy cable used to bring power into the building. It starts at the point where the utility lines are attached to the building. From there it continues through the meter and into the basement or first-floor main distribution panel. This cable consists of two insulated wires and one bare wire, the neutral wire. The neutral wire is composed of many fine wires wrapped spirally around the insulated wires. To use this wire, it must be unwrapped enough to allow the strands to be twisted together to form a wire similar to the hot wires. This cable terminates at the main breaker or fuses.

> CAUTION: Whether the main breaker is tripped or the main fuses are removed, remember that the *incoming* wires and their terminal screws are still *hot* even though the rest of the panel is dead.

A fused panel will have the main fuses on a pull-out block. Pull this out, and you have the fuses in your hand. You may also find yourself in the dark, so be sure either to have a flashlight handy or to get a long extension cord and borrow electricity from a neighbor.

Service entrance cable is approved for supplying electric ranges and electric clothes dryers, but nothing else. For both these fixed appliances, the cable *must* originate in the main distribution panel. Where the house roof line is low, it may be necessary to extend a mast of thinwall through the roof where the eaves meet the outside wall in order to meet the minimum height requirement for the service drop (the wires to the pole). The clearance must be not less than 10 ft above the grade. This mast will continue down and into the top of the meter base. Either cable or thinwall may be used to connect to the main distribution panel inside. *Underground service entrance* (USE) cable has all three wires insulated and is used for direct burial. This means the cable is buried directly in a 24-inch-deep trench for residential wiring. The minimum depth requirement is 12 inches, but 24 inches is safer.

JUNCTION BOXES AND DEVICE BOXES

The NEC requires every device and every wire connection or joint to be within either a metal or plastic enclosure. Therefore, every receptacle, switch, or wire connection must be in a box with a cover. This cover, when there are wire connections inside, must be accessible without damaging the building construction or finish. In other words, the cover must not be plastered or paneled over. Some finished basements or attics have the back side of walls left open. If these areas can be reached through an access panel, then the box can face to that side and meet the NEC requirements.

A variety of electric boxes meet the diverse needs of the industry. Practically every type you will need is available in most hardware stores. These stores and home centers cater to the do-it-yourselfer and homeowner. Wall boxes for receptacles and switches come in different depths and secure them to drywall and paneling by a variety of anchoring methods (see Figs. 4-5 and 4-6). Directions are usually included. Plastic boxes are approved and are cheaper. Since they cannot be grounded, the device *must* be

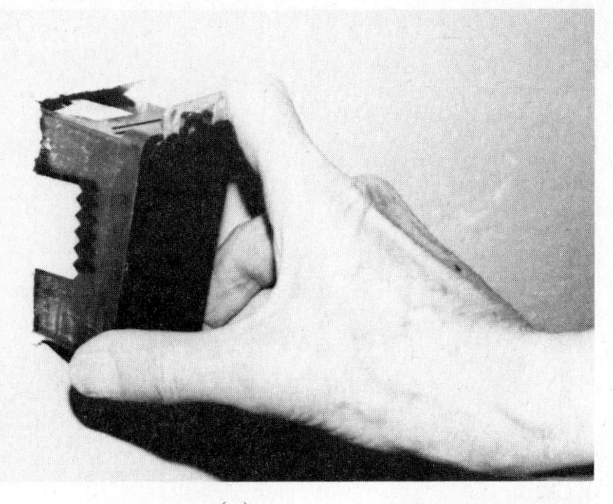

(a)

(b)

FIG. 4-5 (a) Wall box with "spring-out" box holder being inserted into the wall opening. (b) Box in place and secured to wallboard by tightening clamp screw.

grounded. Run the bare ground wire to the green screw on the device. If more than one device is needed, metal boxes can be ganged together; remove the sides, and you can combine two or more.

Ceiling boxes are generally round but can be hexagonal or square (see Fig. 4-7). Square boxes may have a plaster ring that forms a round opening which will be covered by the fixture canopy. (I have two wall fixtures that required very narrow boxes; I was

(*a*) (*b*)

FIG. 4-6 (*a*) A different type of box support. Wings are spread out when screws are tightened, similar to a Molly anchor. (*b*) Box in place. Space around box will have to be sealed with patching plaster.

able to get these special boxes only at a wholesaler.) Brackets that span two joists are used for supporting ceiling boxes in new construction. If the attic is unfloored, these brackets can be used for a light fixture in old work.

The bracket has a sliding box support made of two pieces. Separate the pieces by removing the screw holding them together; mount the box and reassemble the support, now called a *stud*. Slide the box into position, and tighten the stud. Using an adapter called a *hickey* and a short $\frac{3}{8}$-inch pipe nipple, assemble a strong support for a heavy fixture. Most stores have this box and bracket assembled as one item; otherwise, the parts can be bought separately.

> ***Note:*** Only lightweight fixtures should be supported by the furnished strap, which is held by screws driven into the box ears. This is especially important when plastic boxes are used. These ears may have only threads cast in the plastic. Inspect them carefully. It is better to use a metal box.

The *handy box,* for exposed surface work such as basements and garages, is formed in one piece. All its corners are rounded, including the corners on the cover plate, in contrast to standard cover plates, which are too large and have square corners. When you use thinwall tubing with this box, be sure to get the *offset connector*; this saves time and looks neat.

A receptacle for laundry equipment is mounted at switch height, about 46 inches high. Romex is brought down through a section of tubing connected to the box by the offset connector. The metal box and tubing will be grounded by fastening the bare ground wire under a grounding screw turned into a tapped hole in the back of the box. This is the grounding screw hole. You may have to chip a little concrete behind the box to allow for the screw. Because the receptacle, or switch, is in direct contact with the box, there is no need to run a separate wire to the green grounding screw on the device.

Note: The 1984 NEC requires a bushing at the top of the thinwall to protect the Romex from abrasion at that point.

Larger junction boxes are used to make connections to many wires in one enclosure. A fuse or circuit breaker may be relocated to another area, and all the circuit cables will be too short to reach the new location. A junction box will be needed; the short cables enter this box, and *inside* the box additional lengths are spliced on and continue to the new panel location. In this way the NEC standards are obeyed. Most larger boxes have hinged covers with a friction catch.

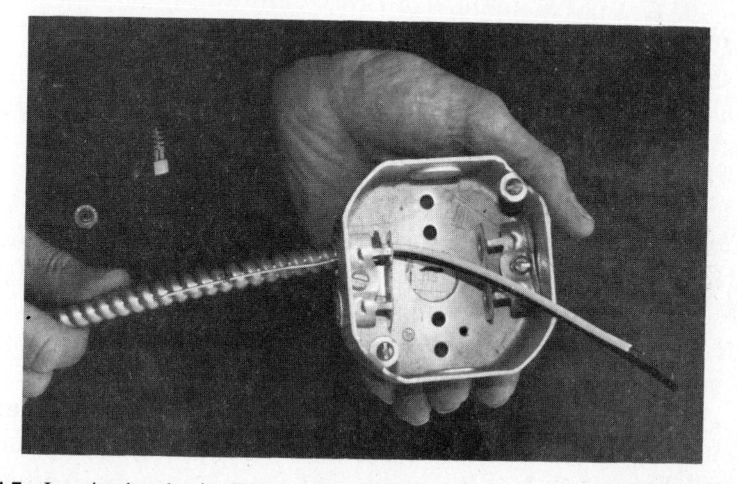

FIG. 4-7 Junction box having BX clamps. Note the stop on the inner end of the clamps to keep fiber BX bushing in place.

ELECTRIC DEVICES

An *electric device* is an electric part that carries current but does not use any current. Included are switches, receptacles, fuses, circuit breakers, attachment plugs, cord connectors, and many more items. Items such as dimmers, electric clock-operated timers, and photocells use a small amount of electricity but are basically devices.

Receptacles

Originally, one of the earliest appliances, the toaster, was plugged into a light socket after the bulb was removed. To avoid twisting the cord, the threaded part of the plug swiveled. On later models, the plug was made in two parts, one with threads and the other attached to the cord and having two brass prongs. The base was screwed into the light socket, then the plug half was inserted into the base half. Around this same time, the wall receptacle was developed that accepted the plug half. The screw half of this type of plug is still sold in hardware stores. Receptacles now have three openings; the half round opening provides a ground for extension cords, appliances, and power tools.

The common receptacle found in dwellings is rated at 15 A–125 V. Because two appliance circuits for use in the kitchen area are rated at 20 A–125 V their receptacles *may* be rated at 20 A–125 V also. If they are, the longer slot will be T-shaped rather than straight. Except for heavier duty they are just the same. They will accept plugs having the special T-shaped prong used on heavy appliances (see Fig. 4-8).

There is an NEC requirement regarding *ungrounded receptacles* (those having only two slots). They are no longer approved. If an ungrounded receptacle must be replaced for any reason, it *must* be replaced with another ungrounded receptacle. If a *grounded* receptacle is installed as a replacement, this receptacle *must* be grounded to a cold-water pipe with No. 14 copper wire. New Romex cable may be used provided it originates at the fuse or breaker panel. Make sure the bare ground wire is connected to the system ground. These special precautions will protect someone who might plug in a grounded power tool, expecting the grounded receptacle

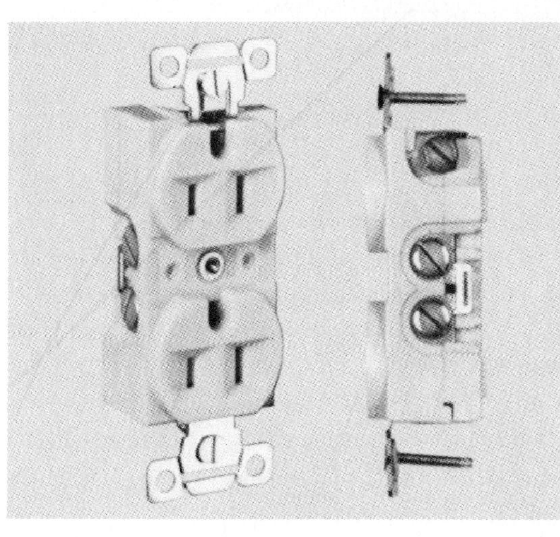

FIG. 4-8 Typical grounded receptacles. (*Courtesy of Bryant Co.*)

to provide a ground for the power tool. If there is in fact no grounding, serious injury could result.

When you are buying electric devices of any type, buy the best you can afford. Many devices now have steel terminal screws, but you should use devices with brass screws. Take a small pocket magnet with you when you shop, and check the screws. A receptacle costing about $2 will be good but not top quality for residential use.

Receptacles are used also for electric ranges and clothes dryers. These special receptacles are large and can be either flush or surface types. Each appliance takes its own style of receptacle. The range receptacle, for example, has three straight slots and is rated at 50 A–250 V. The dryer receptacle has two straight slots and one L-shaped slot; it is rated at 30 A–250 V. These receptacles are the ones found in dwellings. The range and dryer receptacles cost about $5–$7.

Switches

The standard switch costs about $2. Again, check for brass screws. Most switches are the toggle type, although fancy decorative handles are available at extra cost. Actions are snap, silent, and mercury. Mercury action is expensive, so buy the silent type. The toggle switch is always mounted so that the handle is up when the switch is on; this is standard. Three-way switches do not matter, since they can be in either position for on depending on the position of the other switch (see Fig. 4-9).

A line of switches, receptacles, and pilot lights has been developed which enables up to three devices to be installed in a single gang switch box. This line is called the *interchangeable device line* (see Fig. 4-10). A mounting bracket is furnished with three rectangular openings. A device can be mounted in each opening in the bracket and locked in place by a small cam lever. Any three devices can be combined. I recommend using these devices and this bracket *only* when space requirements dictate. The devices are very close to one another and can arc over during thunderstorms and destroy the assembly. An alternative is a combination device, two in one body, such as switch and receptacle or switch and pilot light. These are expensive and, therefore, should be avoided if possible.

Disconnect switches are larger switches enclosed in a metal cabinet with an external on-off handle. These are used for furnaces, water pumps, air conditioning units, and equipment using 240 V. Many have fuses, some do not. Some have only a pull-out fuse block. Sizes depend on the equipment being controlled.

Fuses

The 60-A service entrance panel with its main fuses, range fuses, and four lighting fuses is still in use in many dwellings. Now the

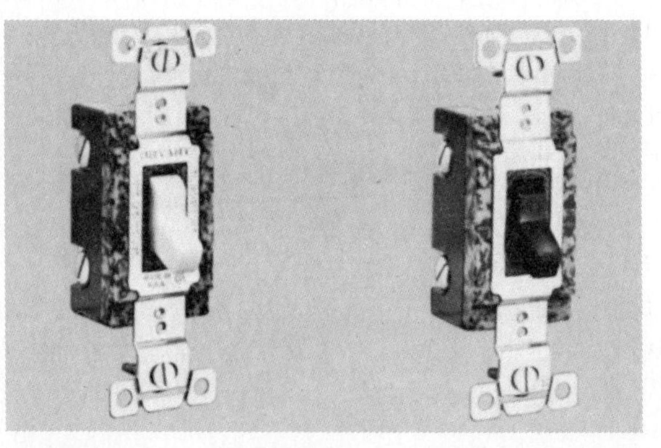

FIG. 4-9 Typical ac switches. (*Courtesy of Bryant Co.*)

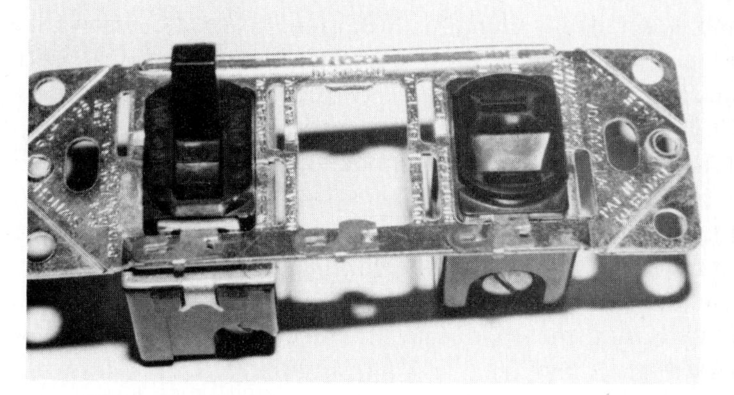

FIG. 4-10 Interchangeable line showing toggle switch and receptacle. Note cam in center opening which locks device in bracket.

minimum size panel allowed in new construction is 100 A. The fuses in this 60-A panel were: two 60-A cartridge fuses, $\frac{3}{4}$ inch in diameter and 3 inches long (the main fuses); two 50-A cartridge range fuses of the same dimensions; and four 15-A plug fuses (these have threads just like light bulbs) for the lighting circuits.

> CAUTION: Fuses rated at 15 A *always* have a hexagonal window; larger-capacity fuses rated up to 30 A have a round window. *Never* replace a fuse with one of a higher rating. Fire could result.

As stated earlier, a new type of fuse was developed because people *did* install larger fuses. Adapters were made to fit the present fuse holders with a barb on the side which made removal impossible. Specially designed fuses were made to fit this adapter; each size fuse would fit only the adapter for that particular size. No substitutions were possible, nor could foil or a coin be used to bridge the fuse.

This system, developed by the Bussman Manufacturing Company, has been adopted and made a part of the NEC. This fuse, known as Type S and manufactured under the brand name Fustat, is designated as *time-delay,* or *lag-type,* fuse. This feature allows a motor to start under load, such as when a washing machine starts its spin cycle, without blowing the fuse. However, the fuse *will* blow under short-circuit conditions; it will hold only for overloads of short duration. Edison-based fuses (the ones with light bulb-type threads) are also available with these features. Bussman Company shortened their name for these lag-type fuses from Fustron to Tron. These time-delay fuses are recognized by their high price and by the stretched spring that shows in the fuse window. The bottom end of the spring is embedded in a small solder pot at the very bottom of the fuse body. On a sustained overload this solder melts, allowing the spring to contract, pulling its end out of the melted solder. This breaks the circuit, but strictly on overload. An overload can be recognized by the contracted spring. If the fuse actually blows, the window of the fuse will be blackened. Regular fuses will give this visual signal also; the opening on a regular fuse usually shows no fuse link in the window after an overload since it has melted and not blown.

Circuit Breakers

Although fuses are one of the best types of overcurrent protection, circuit breakers are being used in almost all dwellings. The disadvantage of a fused panel is that replacement fuses always must be on hand. Fuses always seem to blow on Sunday evening when all the stores are closed and you cannot borrow any. In contrast, breakers are reliable and serve their purpose well. Their advantage is that they can be reset after tripping off.

> CAUTION: When any overcurrent device operates to disconnect a circuit, the cause of such operation must be found and corrected. The usual cause is a defective portable appliance, extension cord, or power tool. Also, a circuit may be overloaded, and another load on the line may cause the overcurrent device to operate.

Breakers are set at the factory for their ratings, usually 15, 20, 25 A, and so on. Any breaker feeding a circuit must be the correct rating for the wire size. Thus a 15-A breaker protects No. 14 copper wire, a 20-A breaker protects No. 12 copper wire, and a 30-A breaker protects No. 10 copper wire.

Since breaker panels are not standardized, a replacement breaker must be purchased that will be compatible with the panel. The breaker package will list acceptable panels.

CAUTION: The new breaker must be the same rating as the old breaker to protect the circuit wiring.

Main Distribution Panel

These items are expensive and will not be needed unless you plan extensive remodeling (see Fig. 4-11). The best way to learn first-hand about the construction and operation of any electric device is to visit a large home-improvement center and spend some time inspecting what is available. Open the panels, work the switches, and check out everything. Since nothing is connected to power, you can safely handle all the devices. Some stores have a whole dwelling system on display for inspection and handling. A good salesperson can answer questions or provide instruction pamphlets.

Support Items

Any wiring materials installed must be held in place by anchors, brackets, screws, and the like. Romex uses staples. These staples may be the standard style or the new style, which is a small plastic bar with a short nail through each end. This style is better since it does not crush the Romex as easily as the other. Neither staple should be driven in so tightly as to crush the cable. Service entrance cable is supported by two-hole metal straps, similar to pipe straps but slightly flattened to conform to the cable. BX uses the metal Romex staples. The staples for BX can be driven tightly, but this is not necessary. To support thinwall, there are two-hole straps or the newer one-hole snap-on strap. This strap is designed to wedge on the thinwall and stay while the screw hole is marked on the wall or ceiling. This feature is handy when you are working alone. For anchoring on masonry walls, $\frac{3}{16}$-inch plastic anchors are inserted in drilled holes, and the item is secured with No. 10 pan head sheet-metal screws. Thinwall tubing and handy boxes are secured by using these anchors also. For drilling in masonry, use a $\frac{3}{16}$-inch carbide masonry drill, which costs about $1.50. Be sure to maintain pressure on the drill bit while drilling. The NEC suggests that Romex running down a masonry wall be protected by thinwall tubing as described earlier. This makes a very neat installation. There are companies that make only brackets and supports for wall and ceiling boxes, most of which are for commercial use.

FIG. 4-11 Residential main and distribution panel combination; main breaker is in place at the top of the panel. (*Courtesy of Bryant Co.*)

Nearly everything you will need can be found in home-improvement centers. As always, look for bargains, close-outs, and reduced prices. Often stores have the ends of coils of Romex, BX, or service entrance cable at reduced prices. Check for the length you need, because you may not be able to return remnants. Also be sure the Romex does have the "with ground" designation printed on the covering. The printing on Romex and entrance cable gives all the information needed, such as number and size of wires, voltage, and type.

CAUTION: Do not buy aluminum wire. Aluminum wire *may* be used for service entrance cable, but I prefer copper.

Low-Voltage Wiring

Low-voltage wiring, such as door chimes, buzzers, pushbuttons, and their transformers are all wired with No. 18 doorbell wire. This is single solid wire covered with waxed cotton insulation. Since the voltage is 12–18 V, this insulation is very thin. The NEC is very clear about not running this wire near line voltage (120-V) wiring. Thus no low-voltage wires may be inside conduit or tubing or any raceway with line voltage circuits. The main reason is the poor insulation on this type of wire. Nor may telephone lines be near line voltage wires. Special junction boxes are provided with a metal divider, which will effectively separate the different voltages. This places phone and receptacles at one location, as in an office.

Doorbell wiring is not done with the exactness of line voltage wiring. It is strung along joists and studs and stapled with "insulated" staples. A fiber insulator covers the top of the staple where it will touch the wire to prevent cutting the thin insulation. Care should be taken to prevent stretching and possible breakage of the wire. Avoid running the wire across the bottom of joists. This invites hanging something on it. Wire connections are made by stripping the insulation and twisting the bared ends. Many times neither wire connectors nor tape is used, and the joint is left bare. Thermostat cable is similar to bell wire but is better constructed. It is composed of up to five wires, all covered with a plastic outer covering. This cable is more carefully installed because of its purpose, which is to control heating and air conditioning equipment.

It is now allowable to install your own telephone wiring. Materials are available where other wiring materials are sold. Wall jacks, phone cable, and all sorts of adapters are manufactured in addition to other brands of telephone instruments. Many phone companies provide printed instructions for the customer to follow. In the home *all* wiring should be concealed within the walls. Phone companies have never concealed wiring in a finished building. You can conceal phone and thermostat wiring in your home; it takes longer, but it is a workmanlike job.

The NEC rules for phone and temperature control wiring generally refer to maintaining separation of *low-voltage* lines from *line* voltage lines. In the phone company instructions and in the NEC, the term *protective devices* refers to the lightning arrestor device located just inside the place where the phone line enters the dwelling. Disregard this, because it is phone company property. Temperature control wiring is referred to as class 2 signaling wiring; it is protected by the furnace circuit fuse or breaker. These wires require the same caution: *keep them away from line voltage wires.*

Television antenna lead-ins and related wiring also are low-voltage and must be kept away from line voltage.

Timers

Timers eliminate the need to turn on and off manually either outdoor or indoor lighting. A timer has a clock motor which turns a 24-hour dial. Metal fingers are attached to this dial, at the desired times, secured by setscrews. These fingers push on levers, which in turn open and close a switch. Each finger is slightly different so as to turn the switch to the on or off position.

All night lighting controlled by a timer still needs to be adjusted periodically to conform to changes in the time of sunset and sunrise. Other timing duties, such as turning interior lighting on and off and lowering and raising building temperatures, can usually be set permanently and will not need adjustment. Electric water heaters often have a timer to turn the heater on for only 2 or 3 hours twice a day, in order to save energy.

Photocells

For exterior security lighting with no need for constant adjustment, **photocells** provide a simple alternative. They are self-contained and respond to darkness by turning the lights on and to daylight by turning them off. Since they are limited in the amount of current they can handle, a large installation can be operated by using a heavy-duty relay controlled by the photocell to switch the lights on or off.

Thermostats

Thermostats are used to control heating and cooling. Combination thermostats have a selector switch for the mode needed, such as "heat" or "cool" or "fan." Some thermostats are strictly electric; new designs are electronic. "Night set-back" thermostats are popular now because of energy conservation. They cost up to $90 depending on the brand. I prefer the Honeywell brand because I had used their products with satisfaction for at least twenty years prior to working for the company. Their products are top-quality. Night set-back thermostats have a timer built into them.

MISCELLANEOUS ITEMS

Given the many different materials and tools on the market, it is possible only to touch upon what may be of interest to the homeowner. Even then, some items may be overlooked. The sections that follow describe the various and sundry items that I have found useful or that I think might come in handy.

Ground Clamps

These clamps are used to connect the system ground to the earth. Usually the connection is made to a cold-water pipe with at least 10 ft of its length buried in the earth. This is a good ground. (See the

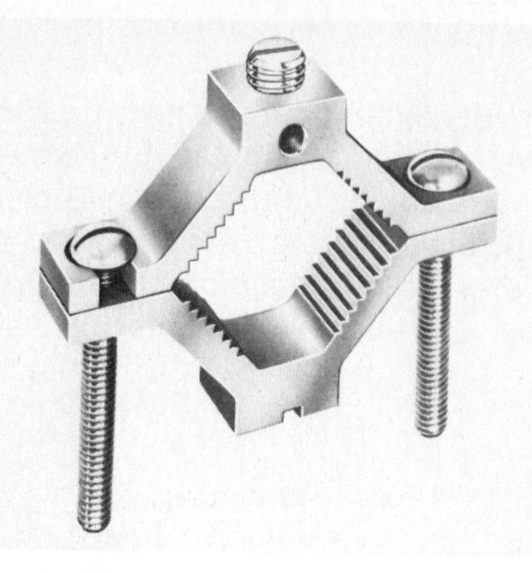

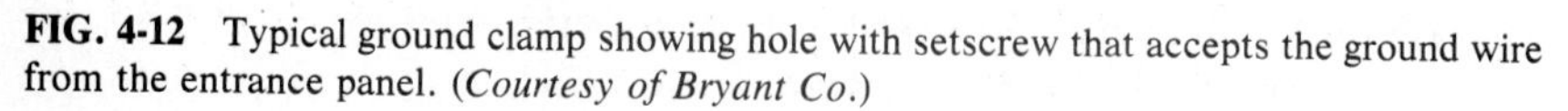

FIG. 4-12 Typical ground clamp showing hole with setscrew that accepts the ground wire from the entrance panel. (*Courtesy of Bryant Co.*)

NEC for detailed requirements and specifications.) These clamps are all similar; they are made in two halves which clamp around the pipe. Two screws are tightened to make a secure contact (see Fig. 4-12). Since the screws are long, different size pipe up to 1 inch can be accommodated. One side has a lug which will accept No. 6 stranded wire. This wire can be bare all the way from the entrance panel; it can be stapled to walls. A setscrew secures this ground wire.

Wire Connectors

The use of wire connectors called **wire nuts** has nearly eliminated soldering of copper wires (see Fig. 4-13). But just as plumbers still have to know how to do lead burning, electricians still must know how to solder copper wire joints. Wire nuts take many forms or styles as manufacturers promote and design their own products; all are excellent. Some are rigid plastic; others are soft nylon. The nylon kind has extended skirts which help cover any bare wire sections. Most electricians wrap tape around the hard plastic wire nuts to eliminate any bare wires; this helps to prevent the nut from becoming loose.

Crimp-on wire nuts are similar to crimp-on wire terminals. They are shaped like a wire nut. The wires are inserted and

FIG. 4-13 There are two types of wire nuts. From the left, the first and third are hard plastic, and the others are soft nylon.

crimped. One drawback to them is that once the nut is crimped, if you find that one wire did not get in with the rest, you must cut the nut off and start again. Standard wire nuts, however, *can* be unscrewed and used again.

Crimp-on Terminals

These terminals are very handy, especially for stranded wire; the strands never seem to stay under terminal screws. Most stores have them, including electronics stores. Check all items mentioned so that you will recognize them again. Straight-through connectors to connect two lengths of wire for greater length are available. Bare the two wire ends about $\frac{3}{16}$ inch; shove them into the connector, and crimp through the nylon; do the same for the other end. This is completely insulated with the nylon. Keep the bared wire end short so that the wire insulation will enter the skirt of the connector.

Split-Bolt Connectors

Large connectors are used to tap onto the power lines at the pole serving your dwelling. They are also used at the connection to your dwelling to connect the service entrance cable to the wires from the pole. The connector is U-shaped; it is slipped over the wire from the pole at a bare spot; the wire from the entrance cable is also put in this slot. When both are in the slot, a nut is screwed onto the two legs of the U which are threaded. When the nut is tightened, the connection is complete. Tape is used at the dwelling end but not always at the pole end.

Electrical Tape

Many types of **electrical tape** are made now, but for the homeowner there is usually the standard vinyl black electrical tape found in the home center stores. When buying this tape, be sure to look for the UL label, which will appear on the wrapper or cover. Some brands not having the UL label may be defective and will either tear or be easily worn through. They may also have no insulating value at all and thus may be dangerous to use and depend upon.

Cable-Wrapping Ties

For wrapping wire bundles and a multitude of other uses, these ties are excellent. The T&B Company was the first to develop these ties and call them Ty-Rap. This name has stuck, as has the name Romex. The tie has a head with a slot with a steel barb inside. The other end of the tie is wrapped around the material to be held, inserted into the head, and pulled through tightly; and the excess is cut off. These ties are used for many purposes never thought of when they were designed, such as tying up tomato plants. Some other brands may be cheaper, but the usual price for Ty-Raps is about $7 per hundred.

Weathertight Outdoor Fixtures and Boxes

For outdoor lighting, cast-aluminum boxes are used. All knockouts in these boxes are threaded to allow rigid conduit or thinwall to be used to provide a watertight connection. These boxes are also known as *raintight*. Receptacles, switches, and light sockets are available to be used with these boxes. Most boxes are rectangular but can be ganged two or more together. The round box accepts a cover having up to three threaded openings for accepting adjustable watertight sockets. All covers are provided with gaskets to make them watertight.

The switch cover has an outside lever to operate the switch inside; a gland seals the lever handle opening. The cover for the receptacle has hinged doors spring-loaded to close when the plug is removed. Different covers are needed for horizontal and vertical box positions because the hinge must always be at the top of the cover.

When thinwall is used with these boxes, the thinwall fitting must be tightened into the threaded opening by using an adjustable wrench or Channellocks. The thinwall is then inserted into the fitting and tightened in place with the setscrew.

Porcelain and Plastic Light Fixtures

These fixtures are mentioned in connection with the installation of lights in unfinished areas of dwellings. They are round, white, glazed porcelain fixtures that fit on a 4-inch octagonal box, or a plaster ring if the ceiling is finished. The three types available are keyless (no switch), pull-chain, and pull-chain with convenience outlet on the side.

In the last few years these fixtures also have been made of plastic, which is extremely durable. Be careful, though, when tightening the mounting screws; if tightened too much, they may break the porcelain base. Some of these fixtures have a groove to accept a shadeholder, if needed. They are convenient and priced reasonably.

Ground Fault Circuit Interrupter

This device was developed to protect people from fatal shock in situations such as when you are using a hair dryer in the bathroom and you touch the metal faucet at the same time. The GFCI is required in bathrooms and also must be used to protect outdoor receptacles and garages, whether attached or detached. This device goes beyond the fuse or breaker and instantaneously disconnects the circuit upon the detection of a very small fault current attempting to pass through a person's body owing to a faulty appliance. Figure 4-14 illustrates a receptacle-type GFCI. This will protect only its own receptacle or may be wired to protect all receptacles "downstream" of it (meaning, a continuation of the circuit

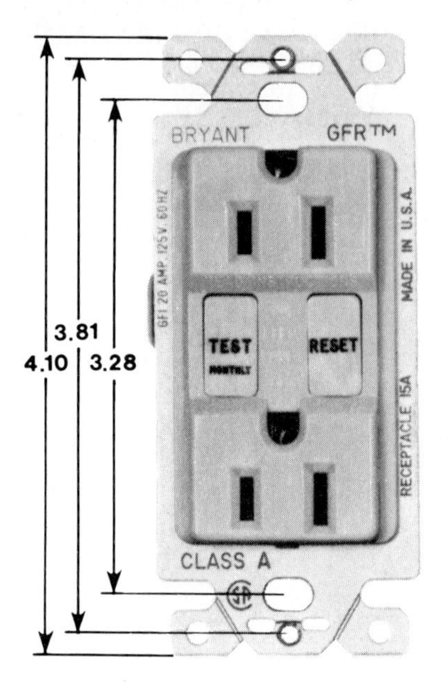

FIG. 4-14 Ground fault circuit interrupter (GFCI). (*Courtesy of Bryant Co.*)

away from the fuse or breaker panel). Most GFCIs come with pigtails rather than terminals. Dimensions are given on the illustration; the depth is as shown, and consideration must be given to the box depth so that the wiring, with wire nuts, can be accommodated. Some GFCIs have a raised plate so that the device can be extended if the box is shallow.

Another GFCI is in the form of a circuit breaker to be installed in the breaker panel. This will protect the entire circuit. A third type is portable so that it can be used on construction sites or in similar situations.

Wiring Circuits Used in a Residence

The number of circuits in a home depends on the size of the building and the extent of the additional equipment installed. Any additional large loads (appliances, power tools, electric space heating, or other large current-consuming equipment) increase the need for larger service entrance equipment. In Chapter 1 we said that this equipment is selected oversize to allow for additional subsequent loads. We begin this chapter by outlining a procedure you can use to determine the present load of your wiring system and decide where you may want to expand your circuits.

EVALUATION OF PRESENT CIRCUITS

The first step in evaluating your present wiring system is to discover which outlets are on each fuse or circuit breaker. One easy way of checking the receptacles (another word for wall plugs) is to buy a light bulb adapter. This device consists of a plastic housing with a light bulb socket on one end and prongs resembling an attachment plug on the other end.

To begin the process of recording the load on each circuit, trip one breaker, or remove one fuse from the panel. Begin at the top of the panel, and do each circuit in sequence; keep a written record of your work to make sure not to miss any circuits. Plug the bulb and adapter into each wall receptacle. It is best to plug the adapter into both top and bottom outlets of the receptacle because sometimes one-half of the receptacle may be on a wall switch. Usually the bulb adapter need not be pushed in all the way to make contact. Also, turn on the ceiling light switch to try that light. All told, this is a tedious job; but once it is finished, you will have a complete record of what part of your home each circuit serves.

Most panels have the circuits numbered. But if yours does not, use a small felt-tip pen to mark numbers in sequence down each side of the panel. Whenever you find a dead receptacle (one with no power), mark a small number corresponding to the circuit number at the panel on the cover plate near the mounting screw. This is not necessary, but it does make a permanent record if you want one.

Use a notebook to chart these circuits and the areas they serve. Take the following example: Circuit 6—small bedroom, north and east wall receptacles; living room, north and west wall receptacles. In this case the shared wall and the north wall are all on circuit 6. You will have to be diligent to find all the outlets on each circuit. Some lights and receptacles may be in the basement or attic while others on the same circuit are in the living room. This is often the case in older houses where outlets have been added to present circuits. Many tract houses will be wired alike only because they were wired by the same person and done at the same time as a group.

Now, depending on your "fixed" appliances (those you don't carry around to use), the distribution panel section will have one main breaker (meaning, a fuse *or* circuit breaker), one range breaker, one clothes dryer breaker, and one water heater breaker.

> ***Note:*** If the heater is on its own meter, the breaker *may* be near that meter.

All the above-mentioned breakers are *double* (joined together) and are 240 V, since they supply 240-V fixed appliances. All the other breakers are 120 V, for they supply lighting and hand appliances.

(An exception to the above 240-V breakers would be 240-V power supplied to a garage or other outbuilding, such as a shop using power tools. This situation might have been provided for by installing a separate small distribution panel mounted beside the large panel.) There is no reason to remove the panel front. But if you want to, first trip the main breaker, which will "kill" (shut off) the power to the whole panel, except the power to the terminals which feed the main breaker from the meter. These will still be *hot*. Be careful when removing the screws from the front and then removing the front. Pull the front directly forward so as not to drag a corner into the wiring inside the cabinet.

It is essential that you move slowly and carefully when you are working around electric panels which are hot or contain certain terminals that are hot. If you are going to remove the panel front, including the door, in such a situation, be sure that you can hold the panel front. It may be heavier than you expect, thus causing it to fall inward. Get help if you have any doubts. It is important to keep any part of the front from falling inward where it could contact hot wires or terminals. If this happens, there will be a short circuit, and you may receive a shock. There is no problem; just do everything carefully.

If you plan to add more appliances or items such as air conditioning or a shop in the basement or garage, estimate the load to be added to the present load. Record this; then calculate the total computed load as set out in the latest NEC. These are the minimum requirements for a dwelling. An example of dwelling calculations is shown in Table 5-1.

TABLE 5-1 Calculations for a Single-Family Dwelling

Fixed appliances	*Estimated load* (kW)
Floor area, 1200 sq ft*	3.6
Kitchen appliance circuits, two 20 A	4.0
Laundry circuit, one 15 A	1.5
Range	12.0
Water heater	2.5
Dishwasher	1.2
Clothes dryer	5.0
Air conditioner, 6 A–240 V	1.4
Total	31.2

* Minimum requirement of 3 W/ft^2.

The sampling dwelling has a floor area of 1200 ft^2 (outside dimensions). The fixed appliances are itemized, and an estimated load is assigned to each one. To convert the air conditioner's amperage rating to watts, use this formula: $6 \times 240 \div 1000 = 1.44$ kW. The estimated loads are totaled (31.2 kW), and the first 10 kW is used at 100 percent. The remaining 21.2 kW is used at 40 percent, or 8.48 kW. Therefore, the calculated load is 10.00 + 8.48 kW, or 18.48 kW (18,480 W). We convert this number to amperes: 18,480 W ÷ 240 V = 77.00 A.

This calculation tells us that the dwelling may be adequately served by a 100-A service. This figure can be broken down into

3600 W ÷ 115 V = 31 A	Two 15-A circuits (or 30 A)
Small-appliance load	Two 20-A circuits (or 40 A)
Laundry load	One 20-A circuit (or 20 A)

Usually you will want more than two 15-A circuits, at least four and perhaps six. The advantage of more than enough circuits is that each circuit carries less load and so will seldom, if ever, be overloaded and trip the breaker or blow the fuse.

Figure 5-1 shows how the current flow must be equal in the two legs (wires) of the 240-V service, even though the neutral wire (N) is needed to carry *any* unbalance of current. The system will operate better, and this may help prevent power outages due to overloads. The worst that could happen with an unbalance of the two legs (wires labeled H in the diagram) is that *one* side of the main breaker could trip off as a result of overload. When the one breaker trips, it causes the other breaker to trip also (because they are linked by the tie bar), and so the place will be dark. In your search for which outlets are on which circuit, you may discover too much load on one leg and decide to reallocate the circuits.

There are simple methods of determining system imbalance at the panel with the proper equipment. I was called to a home once because the furnace wouldn't operate. All other appliances, lights, and receptacles worked fine, but there was no power to the furnace. I traced the trouble to the utility connection at the pole at the rear of the property where a connector had fallen off the wire to the dwelling, allowing this wire to hang free of the supply wire. The furnace was the only load on that leg of the supply. In my own home I once had a similar problem, except that only the clamp

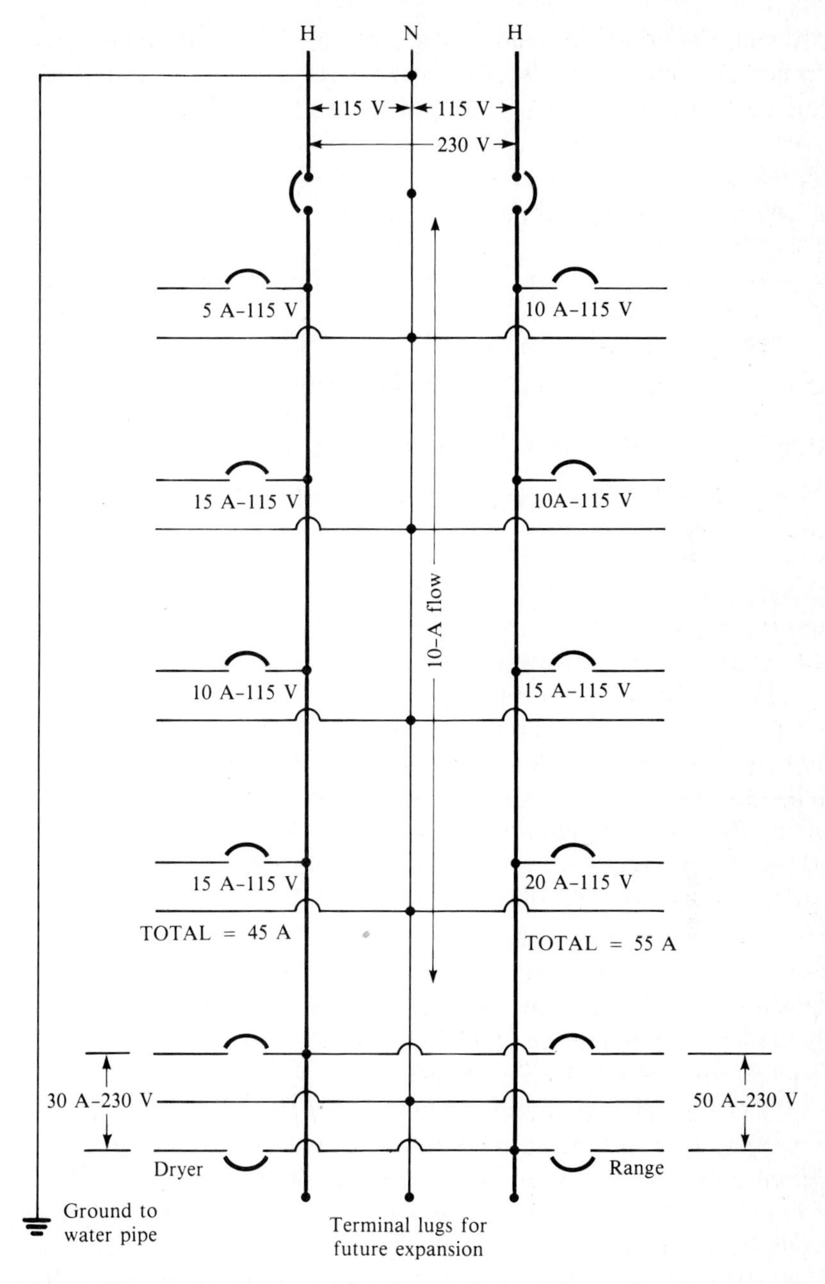

FIG. 5-1 Diagram showing current flow in neutral wire at the service entrance panel.

connector was loose, which caused the lights and appliances to go on and off intermittently. This was worsened by the wind which caused the wires to sway and pull on the loose connection.

> **CAUTION:** Certain appliances may be damaged by going on and off and should be disconnected from the power supply. Refrigerators are especially subject to damage from this condition.

With undersized service entrance equipment, there is the danger of low voltage being supplied to appliances. When the voltage is below normal, the amperage increases. For example, the previous single-family service entrance calculation:

$$18{,}480 \text{ W} \div 240 \text{ V} = 77.00 \text{ A}$$

Note that if the voltage is reduced to 210 V because of undersized equipment, the amperage changes:

$$18{,}480 \text{ W} \div 210 \text{ V} = 88.00 \text{ A}$$

This is an overload of 11.00 A, which may cause breakers to trip.

The interior of a 60-A fuse panel is shown in Fig. 5-2. The capacity of this service entrance equipment is printed on the information sheet pasted on the inside cover. This sheet also includes a directory of circuits to list the area served by each fuse. Figure 5-3 shows a typical wiring diagram for the 60-A service entrance panel shown in Fig. 5-2. A diagram such as this is pasted to the inside cover of the panel. The black wires are the hot wires. The white wires are the neutral, or grounded, wires. The connection of the neutral wire to a ground is not shown because it is not part of the equipment. The connection to ground is "field-installed" (installed by the electrician doing the wiring). The diagram routes all power through the main fuses, including the range circuit. In some methods of wiring, the range fuses are supplied with power directly from the meter and not through the main fuses. This allows the main fuses to carry more load and supply more circuits safely. Your service entrance equipment should be arranged to fit your needs.

Figure 5-4 shows a breaker panel in a condominium unit. The top arrow shows the connection of the main breaker to the meter, as is usual in condominium construction. The panel is rated at 125 A,

FIG. 5-2 Fused service entrance panel, 60-A capacity, showing pull-out block at top of cabinet.

even though the breaker is only rated at 100 A. There are three 20-A circuits (two small appliances and one laundry); three 15-A lighting circuits; one 15-A garbage disposal circuit; one 15-A garage door operator circuit; one 30-A–240-V furnace/air conditioner circuit; one 30-A–240-V clothes dryer circuit; and one 50-A–240-V range circuit in the panel. (Space for more circuits is available when needed.) There is one space that has no breaker or blank cover. A cover should be installed. Figure 5-5 shows an enlarged view of the 50-A range breaker. The arrow points to the tie bar. Figure 5-6 shows the wiring of a typical circuit breaker.

A circuit breaker is designed to quickly break the circuit when a short circuit or ground occurs, just as a fuse would blow. Upon a sustained overload, as when a motor is drawing current in excess of its nameplate rating in amperes, the breaker will trip (break the

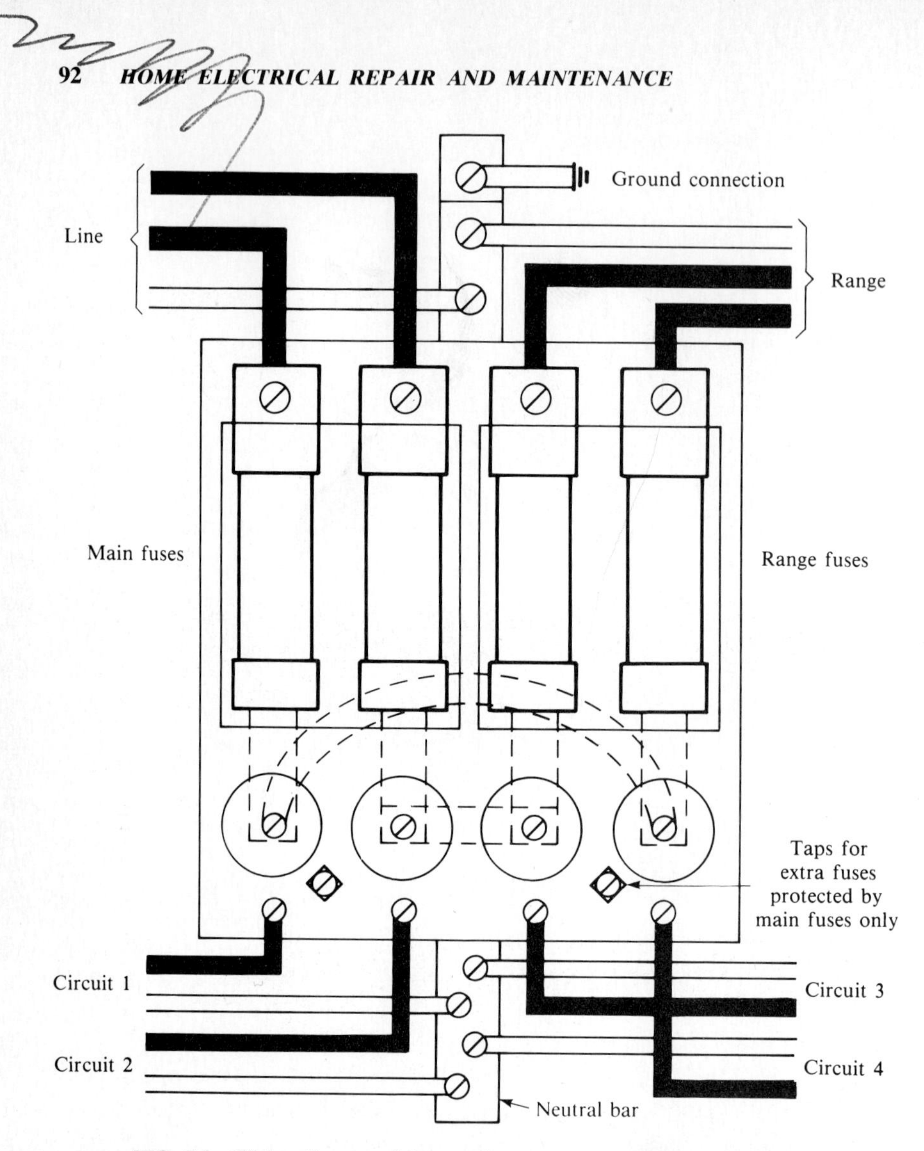

FIG. 5-3 Wiring diagram of the service entrance panel shown in Fig. 5-2.

circuit). This overload *must* last longer than 30 seconds to cause the breaker to trip.

The breaker is designed with an internal element which has a wire or other device that heats a bi-metal strip. This strip bends away from the heat, causing the mechanism to break the circuit. When a motor starts, it will draw current $2\frac{1}{2}$ to 3 times its nameplate rating; this is normal. This does not cause the breaker to trip because the large current draw lasts only 2–3 seconds.

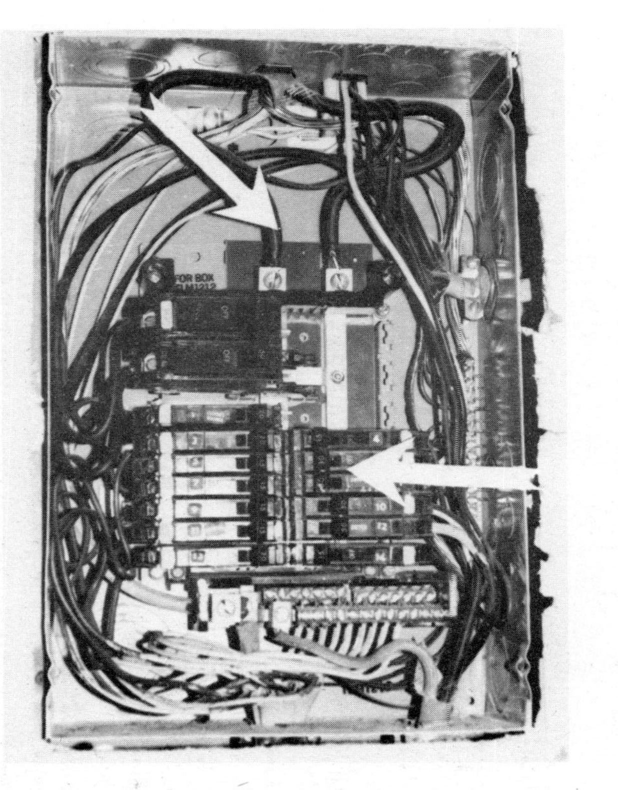

FIG. 5-4 Interior of 100-A breaker panel. Top arrow points to main lugs. Lower arrow points to the double furnace-ac breaker.

FIG. 5-5 Closeup of a range double breaker. Each breaker is 1 inch thick, so together they take up 2 inches of space.

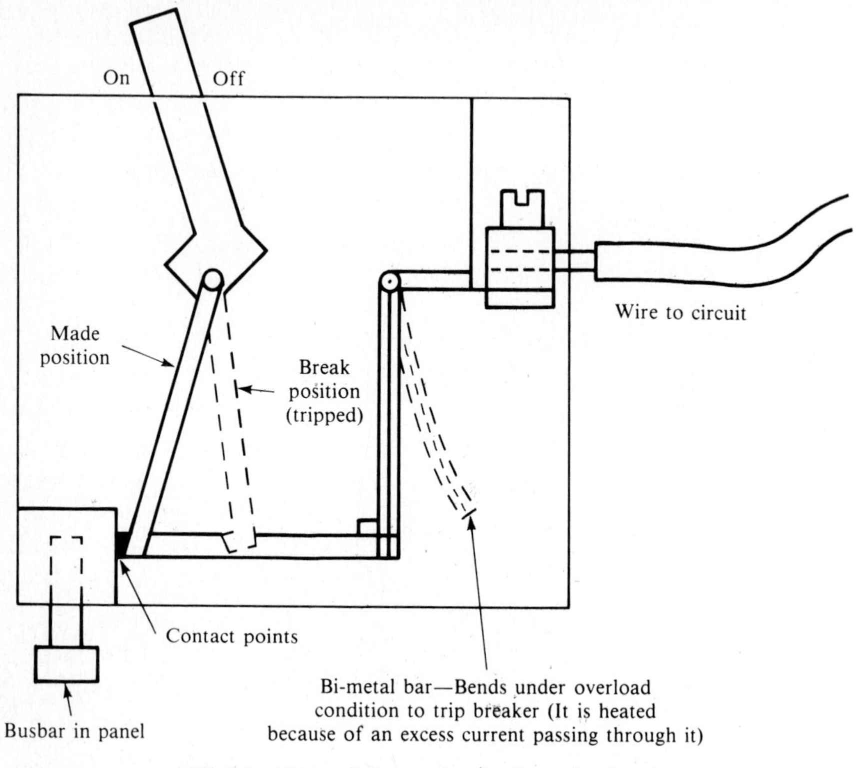

FIG. 5-6 Wiring diagram of a circuit breaker interior.

The Type S fuse operates in much the same way, but must be replaced with a new fuse. The circuit breaker can be reset many times. But remember, constant tripping of a breaker should be investigated to find the cause instead of just resetting the breaker.

Many times the homeowner would like to add more circuits but does not know whether the service equipment will handle the extra load. By using the previous example, it is easy to determine the present load and carry on from there, either adding new equipment or using extra breakers or fuse holders that are already in place.

TYPES OF ADDITIONAL CIRCUITS NEEDED

Additional circuits that may be needed are determined by the type of power desired and the usage. If, for example, you want to add an air conditioning unit, then a 240-V circuit using a 30- or 40-A breaker will be required. From your load calculations of your present panel and your service entrance wiring, you will be able to find out whether you have enough reserve capacity to handle the additional load. If not, then a larger panel and service entrance equipment will have to be installed. If you had only 120-V service in your house, you would have to change your service completely in order to add any 240-V fixed appliances such as an electric range or clothes dryer. Making these changes takes a lot of work, but it's not too difficult.

Additional circuits may be needed in the basement for a workshop or recreation room. It is better to add more circuits than to extend the present ones; thus the chance of overloading is eliminated. Most auxiliary fuse or breaker panels provide space for four or more circuits. There are two-circuit panels available, but you should buy the four-circuit panel for a few dollars more. Existing main distribution panels have lugs provided for connecting an auxiliary panel. These are at the bottom of the panel interior. You can install fuses or breakers for nearly any capacity within the rating of the panel. Wires of the proper size *must* be used to match each fuse or breaker rating. A workshop should have 20-A capacity circuits (more for heavy power tools). The recreation room should have at least one 20-A circuit for a heavy-duty appliance, just as the kitchen does. Evaluate your needs, and add these new loads to the existing calculated load on the service equipment. This will tell you whether the existing equipment has the capacity to handle the added new load.

READING THE ELECTRIC METER

One advantage in learning how to read your electric meter is that you can economize by keeping track of your usage. The electric meter is similar in operation to the hour, minute, and second hands of a watch, taking any three dials on the meter. If the minute hand appears to point to a number, the only way to determine whether the hand is on the number, has not reached the number, or has gone by the number, is to note whether the second hand has passed the 12 o'clock number. The hour and minute hands operate similarly. This is also how the meter dials operate.

There is a 15-day difference between the meter shown in

FIG. 5-7 Reading a watthour meter with a 15-day interval.

Fig. 5-7*a* and the same meter in Fig. 5-7*b*, which shows a usage for that period of 115 kWh. Let's see how we arrived at that reading. Number the dials (1 through 5 from *right* to *left*) on the meter in Fig. 5-7*a*.

- The no. 1 dial has passed 0, but still reads 0.
- The no. 2 dial is on 2 (no. 1 just completed its revolution, which carried the no. 2 dial from 1 to just past 2).
- The no. 3 dial is not yet at 2 because it is only two-tenths of the way around (look at the no. 2 dial again).
- The no. 4 dial reads a little past 2 (no. 3 shows that the no. 4 dial has gone only one-tenth of its next revolution).
- The no. 5 dial still reads 1.

The meter thus reads 12120. The meter has only been in service for 3 years, and with moderate usage dial no. 5 has just started to register. This comes to 367 units per month, average, for a single occupant. Now use what you've learned, and try to read the meter in Fig. 5-7*b*. This difference in readings shows a usage of 57 kWh (12,177 − 12,120 = 57 kWh). Notice the word *kilowatthours* directly under the meter dials in the photograph. The number under the nameplate division reads DE4271888. This is the number given by the utility, Detroit Edison Company, to identify the location and customer. When you read any meter, always use the number *just passed* by the pointer. Study your meter, and the operation of the dials will become clear.

CHAPTER 6

Troubleshooting and Testing the System

Overloads and short circuits are probably the most common difficulties in an electric system. Their sources are different, but their effects are the same: The circuit is shut down by the overcurrent device. But this mechanism is purposely built into the system by selecting the fuse or breaker that will disconnect the power when it becomes more than the system wires can handle.

OVERLOADS

An overload is almost always caused by a condition external to the system. In other words, too many appliances or other current-consuming pieces of equipment are drawing current from the circuit. Sometimes it is a motor which, by itself, draws too much current. An example would be a large table saw in a home workshop. The kitchen and dining areas are apt to have above-normal requirements as well; in fact, the NEC requires that two 20-A "appliance circuits" be installed to supply these areas. This arrangement gives more capacity so that heavy-wattage kitchen appliances can be used simultaneously. Larger-wattage units are now available that were not in previous years. These units need additional capacity to prevent overloads.

SHORT CIRCUITS

Short circuits can be caused by a defect in either the wiring system or the appliance plugged into the system. System defects can be classed as either defects in the materials or devices or defects in the installation of the system. Material defects can be a damaged length of Romex cable or a length of cable with a section of poor insulation between the three wires inside. Devices (anything conducting current but not using current, such as a switch, receptacle, fuse, or breaker) very rarely give you trouble if you buy well-known brands carrying the UL label. Always be sure the UL label is on the device or carton before you buy any electric materials. This is important.

Installation defects are usually the fault of the person doing the actual installation. However, another person working nearby can cause damage unintentionally to already finished work. Furthermore, sometimes outdoor construction jobs are vandalized when they are not fenced in or patrolled. Still, many defects are due to carelessness in installation. So avoid the following pitfalls:

- Romex staples may be driven down too tightly and cut into the insulation. The newer plastic staples are better, having a plastic bar with captive nails on each end; the nails clamp the bar down on the Romex. Even then, don't make them too tight.
- Careful routing of all the wires is important when a switch or receptacle is inserted into a wall box after wiring of the device.
- Since the grounding wire is bare, be careful to keep it away from the hot and ground wires and terminals.
- Making the correct wiring connections is one of the important things to be learned. Accidentally reversing *any* wiring connections will certainly lead to trouble. At the very least, an incorrect connection will blow a fuse or trip the breaker. Therefore, neatness and accuracy are paramount.

If any circuits in your home frequently blow fuses or trip breakers, single them out and find the source of trouble. When breakers trip, they give no indication of the type of problem. The trouble could be an overload or a short circuit. But when a fuse blows, the window of the fuse will be blackened as the result of a short circuit and clear as the result of an overload. An overload will show the fuse window clear because the fuse link has melted and dropped down inside the fuse body. A good fuse has a link shown across its window. There is a notch in the middle of the fuse strip, and this is the narrowest or "weakest" link of the strip. The width of this section determines the fuse rating—a 15-A fuse has a narrower strip than a 20-A fuse. Edison-base fuses are now obsolete; these were fuses with the same base threads as standard light bulbs. Since you could substitute a larger fuse for the original fuse, this type is no longer approved by the NEC. A fuse that is too large for a circuit may allow the wires to become overheated to the point of starting a fire.

Your problem circuit may be caused by a poor connection, such as wires twisted together but not soldered or not having wire connector (wire nut) screwed on the joint. Either situation will cause a high-resistance connection which, in turn, may overheat and damage the insulation on the wires. There is also the possibility of fire.

Other problems may be caused by too many appliances on the circuit or a defective appliance, table lamp, or ceiling fixture. Many ceiling fixtures are tightly enclosed and radiate heat to the wires in the fixture junction box above the ceiling. This can melt the insulation on these wires. Oversize bulbs in this type of fixture can also cause overheating. Use the size bulb specified by the fixture manufacturer. Furthermore, if the breaker or fuse operates when the appliance is plugged in, try this appliance on another circuit. If the appliance causes the fuse to blow or the breaker to trip on this new circuit, then the appliance is defective.

Many times fuses are not screwed in tightly enough, so the fuse heats up and eventually the fuse link inside melts. The fuse will indicate there has been an overload, even though it is neither an overload nor a short circuit.

> ***Note:*** Any connection which is not tight, electrically *and* mechanically, will heat up and cause trouble.

Occasionally check and tighten all screw-type fuses to avoid this occurrence. I have found that in older homes the fiber washer *under* the screw at the bottom of the fuse holder has deteriorated owing to moisture and will conduct electricity even with the fuse removed. Check this if your fuse panel is in a damp location. Remove the fuse and see whether the power to that circuit goes off. This situation is similar to using too large a fuse and gives no protection to the wiring.

Any fuse panel using Edison-base fuses can be converted to the *nontamperable* Type S fuse very easily and at low cost. Fuse holder adapters are available everywhere at a low cost, as are the fuses themselves. Once the adapters are screwed into the Edison-base fuse holders, they cannot be removed. A barb sticking out from the adapter threads will dig into the Edison-base threads when you try to unscrew the adapter.

> CAUTION: Since you cannot remove the adapter once it is installed, be sure to insert the *correct size* adapter. For example, use a 15-A adapter to accept a 15-A fuse (Fustat is the trade name), which will protect a circuit using No. 14 copper wire.

In Fig. 6-1 the lower right holder has the adapter installed.

Low voltage or blinking lights are a problem which may originate either with the utility or in your home. If the light bulbs in fixtures or lamps seem dim, this may indicate low voltage. The utility will usually check the voltage for you and correct it. Utility voltage usually is correct unless lines have been damaged recently. You may be able to persuade the utility to check inside your home at the service entrance panel. This will give you an idea where the problem lies. Small pocket volt-ohmmeters are available for less than $15. They will help you find a poor connection that may be causing low voltage. Or try to rent a voltmeter. Check all receptacles by inserting the probes into the slots where the attachment plug fits. Do not insert the probe in the round opening (the grounding opening); this will not give a true reading. The voltage at receptacles should read 120 V, plus or minus 2–3 V.

In any place where there are terminal screws, there is the possibility that some of or all the screws are too loose. Many times I have been able to tighten some terminal screws one whole turn or more. This is a definite cause of low voltage. Even a slight heating of connections (even the screws on a receptacle or switch) will cause low voltage just when full voltage is needed. In addition, heating of connections expands the parts, making them still looser.

FIG. 6-1 Auxiliary fuse panel with four circuits. Lower-right fuse holder has Fustat adapter installed.

As outlined in Chapter 1, it's safe to inspect and tighten the screws on switches and receptacles when a fuse blows or a breaker trips.

> CAUTION: When you remove a fuse, put it in your pocket. This will prevent someone's replacing it while you are working on the circuit. Furthermore, put a large warning sign on the panel: DO NOT TURN ON—WORKING ON WIRING.

After killing the circuit, remove the cover plate; then remove the top and bottom screws that hold the device in its wall box. Be sure to test again for power (a safety precaution) before touching any wires. Carefully pull the device out of its box; the wires may be stiff, and the box may be crowded. Work carefully, and pull the device out just far enough to tighten the screws. Each terminal screw may have a wire on it. In some cases there may be more than one wire under a screw head, but this is prohibited by the NEC. Investigate this and change the wiring. (This is all explained in

Chapter 7.) Check for correct wiring, such as white wire to white screw and black wire to brass screw. On toggle switches this color matching does not matter. (This is also explained in Chapter 7.)

In the last decade, some receptacles and switches have been designed for *push-in* wiring. There is a small hole directly behind the terminal screws on the plastic block of the device. The insulation is removed from the wire for about ½ inch, and the wire is inserted in this hole (color-matched as before). Once inserted all the way, the wire is secured by a barbed device and cannot be removed. Beside the opening for the wire is another opening for the insertion of a small screwdriver blade. When a screwdriver is inserted and held, the electric wire can be pulled from its opening. Only solid wire may be used in these openings, according to the NEC.

I am not in favor of this connection method because I think that the area between the wire and the holding barb is much too small to provide adequate contact. Most devices have both screws and the clamp mechanism—use the screws. This design is listed by the UL, but many inspectors disagree. There is one advantage to a device having both screws and push-ins; that is, four wires can be installed on both sides of the device. A duplex receptacle that has two terminal screws on each side can have two black wires put under the screws *and* two black wires pushed into the push-in openings. The same number of white wires can also be connected to the white screws and push-ins. Caution must be taken because of the limited space in the wall box. These restrictions are explained later in this chapter.

In many homes, circuits have been extended until the total connected load exceeds the capacity of the circuit. And this reduces the available voltage. This should be part of your initial inspection and evaluation. While you are making this evaluation, look for faulty workmanship, incorrect wiring procedures, and dangerous conditions. Some persons doing wiring do not follow approved procedures as outlined in the NEC. Their attitude is, "If it works, it's OK." Doing electric wiring is an exact procedure, and sloppy work can lead to serious trouble if the work is not installed according to hard-and-fast rules. Remember, fire is always a threat. For instance, the grounded (white) wire *must* be continuous from the service entrance panel to the last outlet. There can be no switch in the grounded wire at *any* point. The system must be grounded at the service entrance panel by a No. 6 standard copper wire (with no splices) from the panel neutral bar to a cold-water pipe (city water pipe). The NEC says there must now be an additional ground such as a **jumper wire** around the water meter. This is installed in case the meter is removed for any reason, so the grounding of the system will be maintained even though the continuity of the water pipe has been broken. One house I purchased did not have any ground wire at all. I had to install a ground wire to have the wiring conform to the NEC and to make it safe for people using the system.

Wiring that is quite old will not have the plastic insulation used nowadays. All wire connections that are wrapped with friction tape over rubber tape should be unwrapped to determine the condition of the joint itself. If the joint seems in poor condition, it is best to cut off the bare wire ends and start over with a new joint. Be sure to clean the bare wires very well. You may need to use sandpaper to make the wires shiny. Soldering of wires other than in the service equipment, in the ground wire, and in all grounding wires is allowed by the NEC, but it is better to use wire connectors (wire nuts). The soldering of joints is shown in Fig. 6-2. Plastic electrical tape must be used if the joint is soldered.

Any type of electric connection must be both electrically and mechanically secure. In addition, any twisted wire joint, be it a pigtail or tap joint, must be soldered or have a device (either a wire nut or a clamp connector) to secure the joint mechanically.

Wall and ceiling boxes, whether plastic or steel, have a maximum capacity, or "fill," of wires allowed in the box, depending on the box size. The standard $3 \times 2 \times 2\frac{1}{4}$ inch switch or receptacle box has a capacity of 10.5 in^3 and will hold five No. 14 or four No. 12 wires. A No. 14 wire requires 2 in^3 of space, and a No. 12 wire requires 2.25 in^3 of space. Tables stating these requirements are found in the NEC, Sec. 370-6.

If you have many wires coming into a junction box, the largest one in this line of boxes is the $4\frac{11}{16} \times 4\frac{11}{16} \times 2\frac{1}{8}$ inch box. This box has a capacity of 42.0 in^3 and can hold twenty-one No. 14 or eighteen No. 12 wires. Be sure to adhere to these capacities, as overheating the wires may damage the insulation. If this is just a junction box, remember that the box must have a cover on it and must be accessible without disturbing the building construction. A lift-up ceiling can have such a junction box above it because the ceiling panel can be replaced easily.

All grounding conductors, those used for grounding the metal

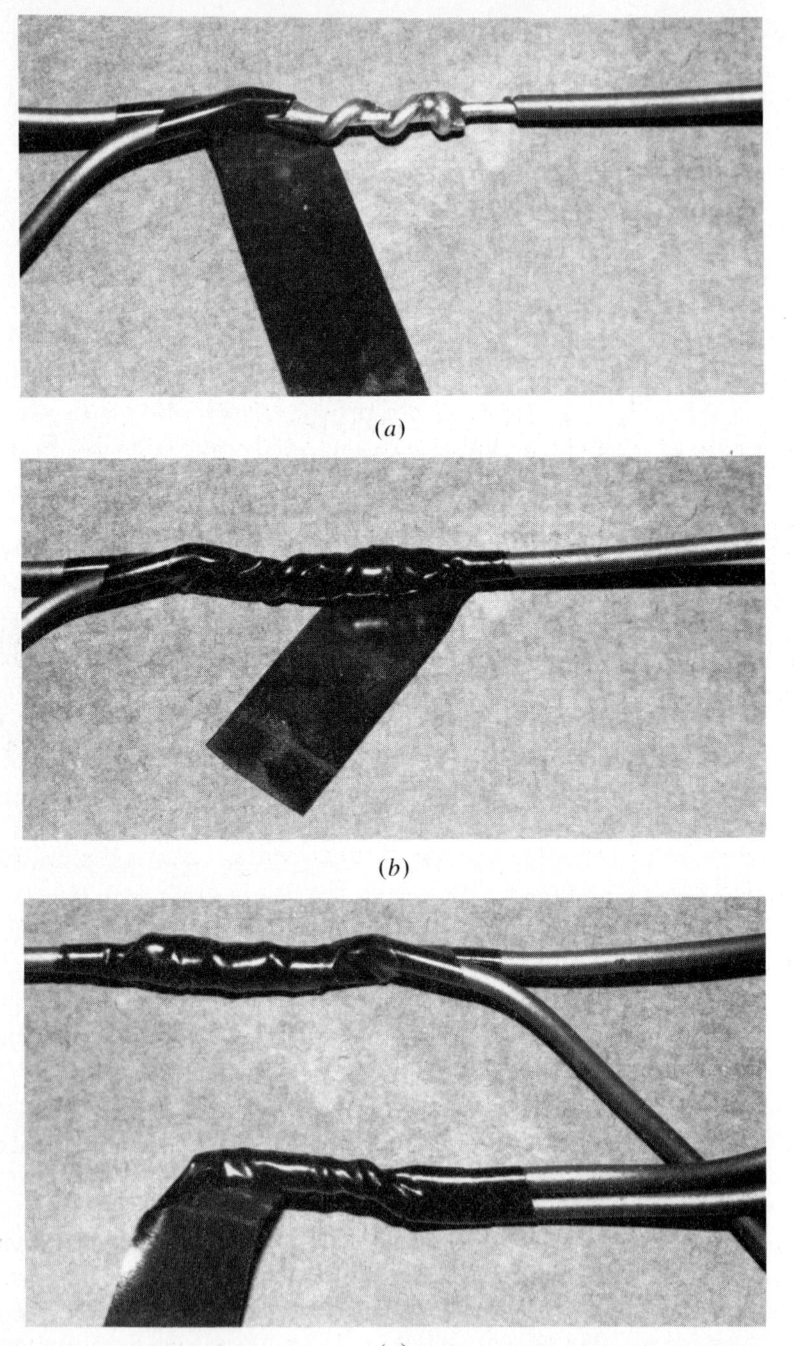

(*a*)

(*b*)

(*c*)

Fig. 6-2

(*d*)

FIG. 6-2 Once a wire joint is properly soldered, it must be taped as shown. (*a*) Tape is started to the left of the bared and soldered section. (*b*) Taping is finished by being wrapped back on itself for added insulation thickness. (*c*) A pigtail (end) type of soldered joint is taped in a similar fashion. The tape is partially folded back to cover the joint end and then wrapped back on itself to add thickness to insulation. (*d*) A second layer of tape is added to secure the connection.

parts of the devices and the boxes themselves, are counted as only *one* conductor.

If you want to learn to solder copper wires together, you will need:

1. *Rosin-core* wire solder (acid-core solder is prohibited)
2. A fairly heavy-duty soldering iron or a trigger-type soldering gun
3. Plastic electrical tape (The electrical tape must be designated as such and have the UL label; it will be black.)

Since the modern plastic insulation does not adhere to the wire itself, the wire is always clean when the insulation is removed. Formerly, the old rubber-covered wire was very hard to get free of all the particles of rubber—it took a lot of scraping. Furthermore, wire nuts were not yet on the market, and alcohol torches were used in place of soldering irons.

CAUTION: Never leave an area immediately after using an open flame. Stay around at least another 30 minutes, or have someone else check for the rekindling of any sparks which went unnoticed. These could develop into a major fire.

Having stripped the wires of $\frac{3}{4}$-inch of insulation, twist them together by hand to start, then use electrician's side cutters to make a tight twist. Next, cut the joint to about $\frac{1}{2}$ inch in length. The joint should now be electrically and mechanically secure—now solder and tape the joint.

If your overall inspection and evaluation showed any defective or poorly installed wiring, you may have a problem circuit which you should take care of as soon as possible. Usually, the problem is that there are too many outlets connected to one circuit—circuits were just added without sufficient capacity to back them up. The best solution for an overload is to split the circuit into two circuits. The recommended way to do this would be to divide the circuit in half, but this may not always be possible. If the circuit wiring is accessible in the basement or attic, then you can easily make the division (see Fig. 6-3).

It will be necessary to have a new circuit feed to come from the distribution panel, since all circuits originate there. If the panel has no unused fuse or breaker spaces, it will be necessary to buy and install an auxiliary panel and mount it alongside the original panel. Breaker panels will usually have the extra space, but auxiliary breaker panels are available with space for two or more breakers depending on your needs. The auxiliary panel should match the main panel, either fused or breakers. Prices start at about $14 for the auxiliary cabinet, and the breakers cost about $7 each. Shop around in hardware stores and home-improvement centers for the best prices.

Continuing with the division of the circuit, look for a suitable receptacle or other junction point to terminate the first half of the circuit. For example, after removing the fuse or tripping the breaker, remove the receptacle that you have chosen as a termination point and pull it from its wall box. This receptacle should have one cable feeding from the panel and one cable continuing on to feed the rest of the circuit. Thus, two white wires and two black wires will be attached to the receptacle.

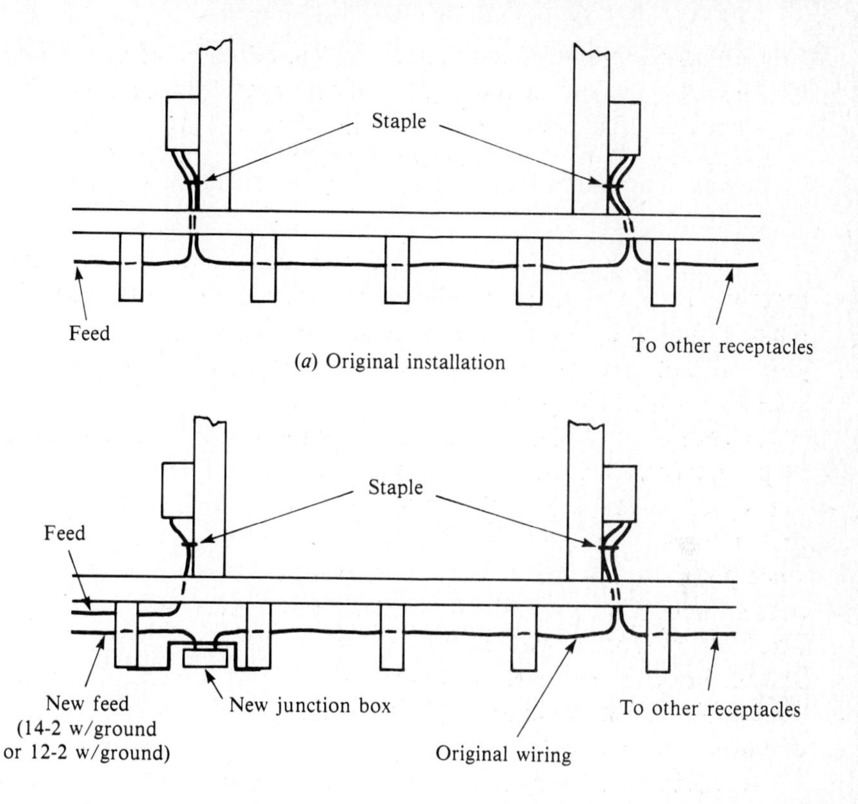

FIG. 6-3 (*a*) Diagram of a circuit feeding several receptacles in sequence. (*b*) Diagram of a circuit terminated at a receptacle to the left. The balance of the circuit is fed from a new line brought from the panel, using perhaps an unused circuit.

Guessing which cable feeds to the receptacle from the panel (with the hope that it will also lead to the second half of the circuit), with the current off remove the two wires, one white and one black, that go to the other cable from the receptacle. Carefully disconnect the wires from under the terminal screws. If they are in the push-in holes, use a small screwdriver to push where it says "release" on the receptacle back, at the same time pulling on the wire (you will need three hands). It is sometimes very difficult to pull the wires from these push-ins. Usually the difficult ones are the cheapest. Now insert your light bulb adapter tester in the loosely hanging receptacle. Make sure that all wires are separated from the receptacle and from each other; then go back and restore power.

> CAUTION: Know where small children are as you restore power. The wires from the panel to the receptacle are hot and exposed.

If the bulb of the tester lights, the feed from the panel is still connected to the receptacle—this is the result you wanted. If the bulb doesn't light, check the tester in another circuit to be sure the bulb is still good; then insert it again in the loose receptacle to double-check. If the bulb still doesn't light, turn off the power again, reattach the disconnected wires (where they were before), and remove the connected wires. *Be sure the power is turned off when you are making the change.* Check again for power to the receptacle. If the tester bulb *still* doesn't light, you can change your approach. It is quite certain that one of the two cables feeds the receptacle since the receptacle was hot before you started. Recheck your work because it's easy to mix up the wires. After loosening the cable clamp inside the box, try to pull the cable that continues on from the box. In the center of the cable clamp there's a screw which can clamp two cables if they both enter at the same end of the box. You may be able to pull the cable from the basement or attic. One of the two cables will be dead when disconnected from the receptacle—this is the one to pull from the wall box and into the basement or attic.

Make absolutely certain that the cable you disconnected is the correct one. Have a helper shine a flashlight into the box as you wiggle and pull on the cable from the basement or attic. If the helper sees movement in the box, stop and then wiggle again. If there is movement again, you have the right cable. Do not jerk on the cable too hard, or you may break the wires. Sometimes the staple is not tight, and the cable comes out easily. Roll up a ball of electrical tape to stuff in the opening left by the cable in the box; then tighten the cable clamp again to hold the remaining cable. Push the receptacle back in the wall, carefully bending and routing its wires so as to reduce interference with the receptacle. Install the mounting bolts, and replace the cover. Turn the power back on, and test the half-circuit that is still connected.

If you have successfully loosened and pulled free the cable

leading to the second half of the circuit from the receptacle box and receptacle, you can install a junction box at the point where the cable ends in the basement or attic. Use wood screws to mount the box. The original cable will usually be No. 14 wire, with or without a grounding wire included. If the old cable had no grounding, it may be difficult to install new grounding cable. (Wiring methods are explained later in the book.) The old cable can be connected with the new grounded cable. The grounding (bare) wire will not be used, but do not cut if off; connect it to the box itself, if the box is metal, for possible future use. There is a tapped hole in the back of the box designed specifically to accept a grounding screw; 10-32 threaded, hex-head, green grounding screws are available. Since this is not new work, the NEC is not violated.

Connect the new cable to the old cable in the junction box, white wire to white wire and black to black. If the old cable is the grounding type, also connect the grounding wires together in addition to a bare grounding wire to the new junction box, if it is metal. For a plastic box, connect the grounding wires with a wire nut, and push them back into the box. All connections, including the bare grounding wires, must be completed by using wire connectors (wire nuts).

Many older homes were wired with the minimum capacity, even as small as 30-A service equipment and two circuits of 15 A each. These dwellings with inadequate wiring may still be found in rural areas and small towns. There are probably even some in central cities. If this is the case with your house, consider a complete rewiring job. This entails the abandonment of all the old wiring and installation of all new wiring. This is a large undertaking, indeed, but it will increase the value of your house. And the complete job can be done by the do-it-yourselfer with adequate knowledge and instructions. Otherwise, the work can be contracted out in part, such as the installation of the service entrance equipment and the distribution panel, with the balance of the wiring easily done by the homeowner.

CHAPTER 7

Repairs and Additions

Simple repairs of the home wiring system are easily made provided you follow directions and work carefully. As mentioned before, you must work accurately and be precise. Always make sure the power is off when you are working on any wiring in the house. You have learned how to test for power (see Fig. 7-1) and have been advised to test at least twice for power. The cardinal rule is that a wire is assumed to be hot (live) unless the full length of the wire is visible, and that includes both ends! Therefore, when working on concealed wiring, such as the kind found in the home, be doubly sure the parts you are working on are de-energized.

REPLACING RECEPTACLES

Receptacles are probably the easiest devices to replace, if only because you can sit down while working on them. When you discover a receptacle in poor condition, such as a plug that will not stay in the receptacle or a plug that must be wiggled to get power from the receptacle, it is wise to replace the defective receptacle.

FIG. 7-1 A device called a plug-in socket adapter is used for testing receptacles for power.

Before starting work, purchase the replacement receptacle. Don't waste time installing those used devices you may have on hand. They may fail soon after installation, forcing you to do the work again. New receptacles are made in many grades of quality, from the bargain-basement (49-cent sale-priced) quality to the seemingly indestructible hospital grade. The top specification grade has a wholesale price of about $3.

The new receptacle should be identical to the original. If it is ungrounded, replace it with ungrounded. If it is grounded (identified by two slots plus one round hole for the grounding prong), replace it with grounded. As always, first turn off the power and then check at the receptacle for power. Next, take off the cover plate. Again, check for power. (Always check for power twice.) Remove the two receptacle-holding screws, and pull the receptacle from its wall box. Some circuits are wired with No. 12 wire, which is quite stiff. You may be pulling on four of these No. 12 wires. Just pull steadily. Check to see that the white wires are connected to the white screws and the black wires to the black screws. In some cases the break-off tab between the brass screws has been removed to separate the circuits or for other reasons. Likewise, occasionally the break-off tab between the white terminal screws is removed. Check this out so that the new device is identical to the old device.

Assuming you have four wires, two white and two black, back out one screw at a time and remove its wire. Use needlenose pliers to open the wire loop enough to remove it from the screw. A small screwdriver also helps. Remove each wire in turn, spreading them out toward the four corners of the box to help you remember to return them to the correct screws on the new receptacle. Most receptacles have the screws already backed out. Reattach the wires color to color, each in turn, and tighten snugly. Do not use excessive force. Very snug is fine.

> ***Note:*** If the wires from the wall box are quite long, cut them off at the screws on the old receptacle, strip them, and form new open loops so that they will fit under the screws. Close the loops with needlenose pliers. Opening and closing the copper wire loops may cause them to crack and break. New loops are better if you have the extra wire length to work with.

Many devices have back-wiring construction. In that case, the wires may be straight and inserted into a special hole with a holding clamp inside, which secures the wire. Another hole next to this wire hole is for the release of the wire, if necessary; a small screwdriver can be pushed into the release opening. On other devices, the straight wire has to be pushed in behind the terminal screw and tightened in this position, thus making a tight connection.

Figure 7-2 shows a *nongrounded* receptacle and how to check that the metal wall box is actually grounded. The tester being used is a neon bulb pocket tester. If it is determined that the box is grounded, then a grounded receptacle can be installed. The photo-

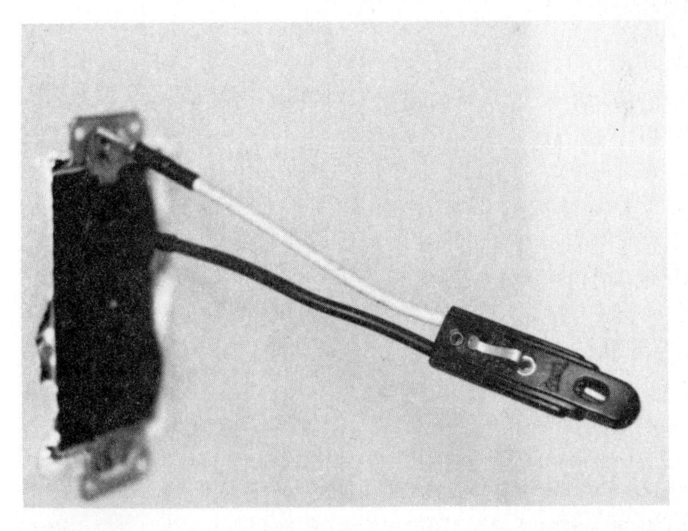

FIG. 7-2 Testing for a *grounded* metal box so that the installed receptacle will be grounded.

graphs show a special mock-up, illustrating how the grounding wire is attached to the box by a grounding clip. Another wire is attached to the grounding screw on the receptacle mounting frame. Because wall boxes are often recessed into the wall $\frac{1}{4}$ inch or more, the mounting screws going into the metal box do not allow the receptacle mounting frame to contact the box metal directly and so do not make good contact. The plaster ears will hold the receptacle flush with the wall, but the receptacle frame is not in positive contact with the metal of the box. This is why you *must* connect the grounding wire to the receptacle grounding screw (green). When two cables enter the wall box (the circuit continues to other receptacles), you will have a wire connector (wire nut) connecting *four* bare wires:

1. From the entering cable
2. From the leaving cable
3. From the metal box clip
4. From the receptacle grounding screw

Note: When you are using wire connectors, first make sure that the connector has the capacity for the number and size of wires given. Also, after screwing on the connector, pull on each wire in turn to see that they are all clamped tightly in the connector. Many times one wire will appear to be gripped by the connector, but comes free when pulled. This can cause a short circuit when the loose wire is one of the current-carrying wires.

There are now receptacles that have a special spring on one mounting screw; this eliminates the need to use a receptacle grounding screw to ground it.

CAUTION: The wall box *must* be made of metal in order to use this method; if the box is plastic, the receptacle grounding screw must be connected to the system ground (the bare grounding wire in the nonmetallic cable).

If you determine that there is *no* system ground at the wall box, you can either install a nongrounding receptacle as a direct replacement or rewire the circuit with nonmetallic cable with ground and use a grounded receptacle.

REPLACING A WALL SWITCH

Wall switches are available with various styles of operating handles. The standard toggle-handle switch is most popular. Most other styles require a special plate because of their design; some have a rocker handle which is much larger than a toggle handle. A third style features nearly the entire cover plate as the handle. The style of handle has no bearing on the switch mechanism; many switches are not the quiet type. Switches are also available in various grades or qualities of construction and rating usage.

> ***Note:*** Lesser-quality devices have *steel* terminal screws. Better-quality devices have *brass* terminal screws; also, the terminal screws are larger. When you buy these devices, take along a small magnet and check the screws. Reject those with steel screws. The better devices cost more but are of better quality and will last much longer. (This advice also applies to receptacles.)

There are three designations of switches: (1) **quiet,** (2) **mercury,** and (3) **snap.** Most switches (except the snap) operate quietly. Many bargain switches do not have a positive on or off position after a period of use. The commonly used grades of quality are *standard,* or *residential, grade* (the lesser quality) and *specification grade* (the better quality). The standard grade, priced at about $2, is found in home-improvement centers and hardware stores. The specification grade is sold at electrical wholesalers for about $3. In any case, the better switch or receptacle will have a larger, more sturdy body and the look of quality. Certain switches will have the green hex grounding screw, and some will not. The switch yoke should be grounded if the screw is present. This screw on the *receptacle* is important because it provides grounding for any portable equipment plugged into the receptacle through the third prong of the equipment attachment plug and the grounding opening on the receptacle.

> CAUTION: *Never* cut off the grounding prong of an attachment plug; this is for your safety.

Grounding continuity is maintained by the *bare* ground wire of the nonmetallic sheathed cable being connected to the receptacle grounding screw.

The actual replacement of a switch is similar in procedure to that of the receptacle. Switches (except three-way switches) are installed so that the up position of the handle is the on position of the switch. Mercury switches have the word *top* or *up* stamped onto the top end of the yoke since the switch will not operate in any other position.

To replace a switch, begin by turning off the power. Next, remove the faceplate and the screws holding the switch in place.

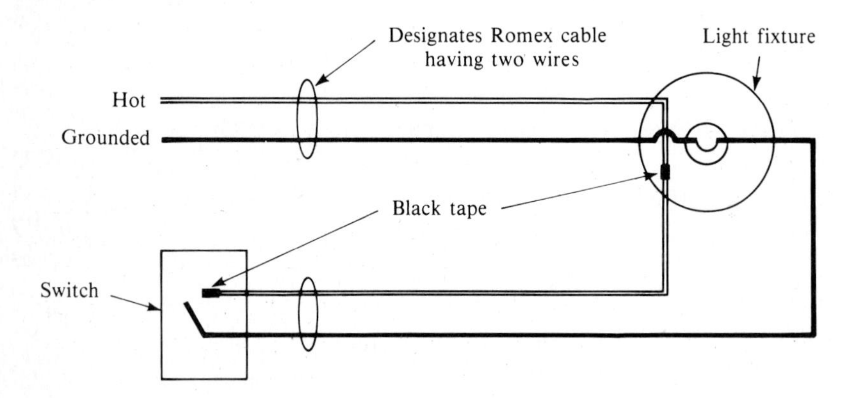

(*a*) Romex cable is used for the switch loop

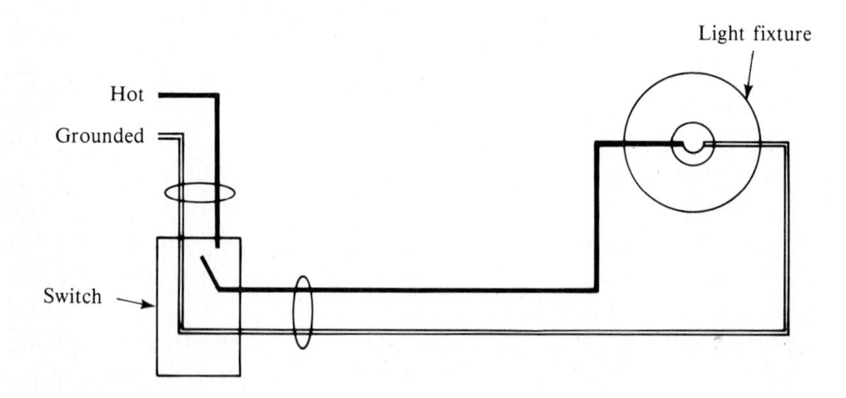

(*b*) Romex cable feeds power to the light through the switch (wire color is correct)

FIG. 7-3 Wiring arrangement for a toggle switch.

Gently pull the switch from the wall box far enough to be able to disconnect the wires. As before, if the length of the wires will allow it, cut them at the terminal screws instead of undoing them. There may be one white wire and one black wire. This doesn't matter, since both black and white are considered black (see Fig. 7-3). (The white wire is the ground, and the black wire is hot. This is according to the NEC. Figure 7-3*a* is the *only* exception allowed by the NEC.) Check the new switch to see that up means on, and then connect the wires, making new bared ends if necessary. Tighten the screws and place the new switch in the wall box; fold the wires back in carefully. Sometimes it is wise to try the switch "hot" before you go to the trouble of replacing it in the wall box. The correct way to do this is to turn the switch to on before restoring the power. When you turn on the power, the light controlled by the switch should be lighted. If so, turn off the power, replace the switch in the wall box, and install the switch plate.

Three-way switches perform the same function—turning lights on and off. The difference is that the control is from either of two locations. There are different ways to wire three-way switches, depending on whether the light being controlled is *between* the two switches or *beyond* the two switches. Figure 7-4 illustrates these differences. The switch will have three terminal screws, one dark bronze and two brass color screws. The bronze screw is the one to which the incoming black (hot) wire is connected. The two brass screws connect the two wires, black and red (called *travelers*). Since *three* insulated wires are needed between the two three-way switches, you will need to buy No. 14-3 Romex with ground. From the second switch to the light fixture, use No. 14-2 wire (Romex) with ground. If you are only replacing the switch, the wiring is already in place. At times you will want to rewire in order to have two locations (switches) for control of a light fixture such as a stairway area. Four-way switches are available to control lighting from more than two locations. They cost more than double the price of a three-way switch and aren't usually needed in a residence.

ADDING A NEW RECEPTACLE

Adding a new receptacle to an existing circuit is a routine procedure. These directions assume that you will be running the wiring

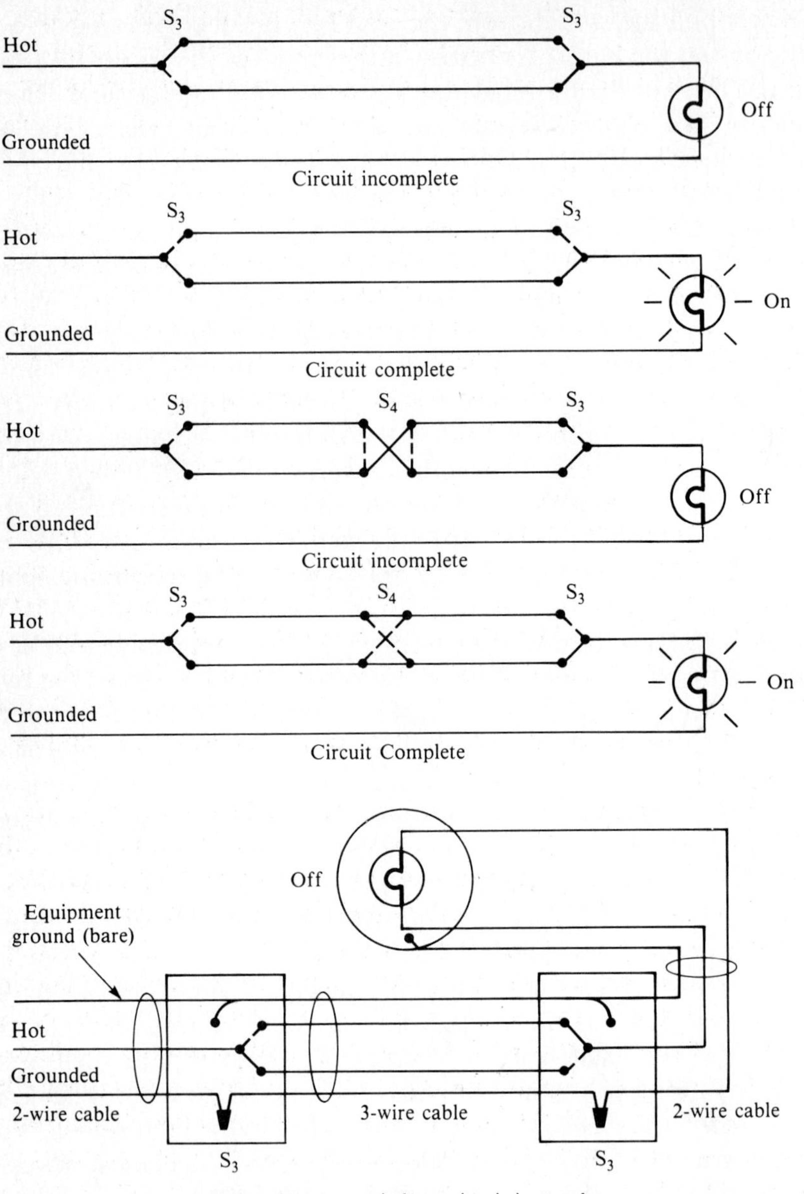

FIG. 7-4 Wiring and operation of three- and four-way switches. S_3 denotes a three-way switch; S_4 denotes a four-way switch.

on the basement ceiling and that the basement ceiling is unfinished. First, determine the location of the new receptacle on the room wall space. Pry out the shoe molding (quarter-round) below where the receptacle is to be located. Drive a 6-inch spike at an angle down at the corner where the floor and baseboard meet (if you have a long twist drill, you can use that). This will show on the basement ceiling. When you have located the spike or drill point, drill a $\frac{5}{8}$-inch hole in the center of the wall space, measured carefully, so as not to drill up through the finished flooring on either side of the hollow wall space. Use a spade bit to make the $\frac{5}{8}$-inch hole up into the wall. With a straightened wire coat hanger, probe up into the hole you've just drilled to check for obstructions. If none are found, you can cut the wall opening in the drywall or lath and plaster. Figure 7-5 shows how to do this easily and carefully. Try to locate the wall box next to a stud so that the box can be anchored solidly. To locate the stud, use a magnetic stud locater or knock on the drywall. (A hollow sound indicates a hollow space; a more solid sound indicates a stud.) By using a bent piece of wire coat hanger, again you may be able to locate the stud by probing from the basement through the $\frac{5}{8}$-inch hole just drilled. Make all measurements carefully, and *think* before you cut. Punch a $\frac{1}{8}$-inch hole in the drywall; probe with an L-shaped wire to locate the stud from this point.

When you find the stud, hold up the wall box at the right height and mark its outline for cutting. Match the height with those in the area for uniformity of appearance. Cut the drywall with a hacksaw blade removed from its frame. Tape one end for a handle. Start cutting with the blade almost parallel to the wall surface. When the blade penetrates, saw more at right angles to the surface. Do this on all four sides. There is no need to drill any starter holes in the drywall. Make the box hole as snug as possible without binding (it should be an easy fit with no gaps). Anchor the box to the stud by drilling two holes in the box side. Metal boxes may already have the holes. Since you will be using the screwdriver at an angle, the holes cannot be too far into the box, or else you will not be able to reach them to drive the screws. About 1 inch is the maximum depth. Pull the cable in from the basement and out through the box opening in the drywall; also draw it in through the wall box, leaving 8 inches of length inside the box for connections. Anchor the cable to the box if the box has anchors. Curl the end of the cable outside

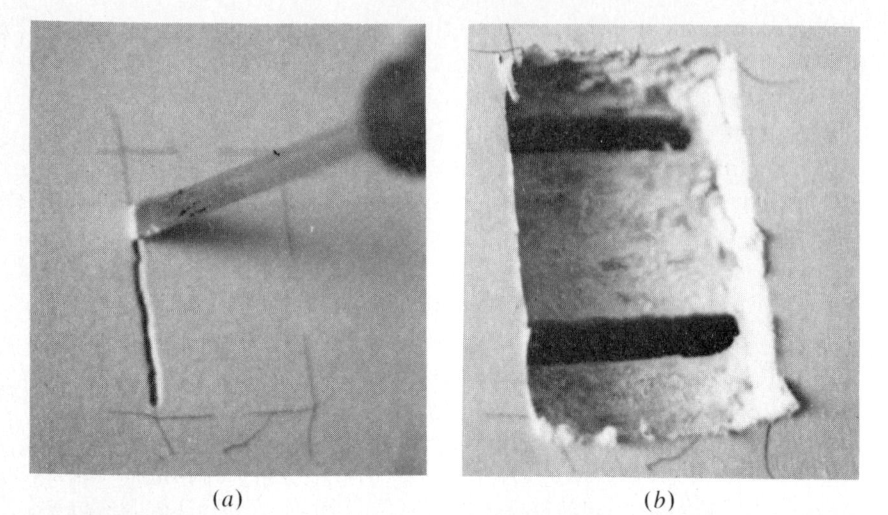

(*a*) (*b*)

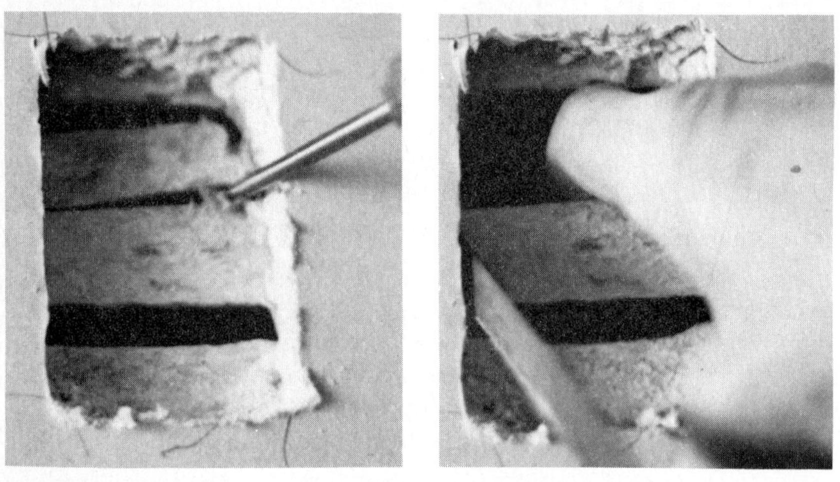

(*c*) (*d*)

FIG. 7-5 Cutting a hole for a wall box. (*a*) Plaster is chipped out by tapping on the handle of a screwdriver with electrician's side cutters. (*b*) Plaster is chipped out down to the wood lath. (*c*) Lath has been partially sawed on each side. A screwdriver is shown splitting a piece of lath. (*d*) Finger holds lath for final cutting. This prevents the lath from breaking away from the plaster.

the box to prevent losing it down into the basement, and insert the box into the opening. While holding the box flush with the drywall surface, drive screws into the stud through the box side where you drilled the holes. There are patented box supports which you can use when the receptacle box is not located so that you can anchor it against a stud. Other boxes have special screw-type clamps similar in action to Molly anchors. Tightening screws on each side of the box spreads wings behind the drywall, thus holding the box in place (see Fig. 7-6).

When buying Romex, look for sales in hardware stores and home-improvement centers. Many times there are good buys on supplies that you may need. Romex is sold in lengths of 25 and 50 ft.

> **CAUTION:** The size wire used in a circuit is directly related to the rating of the fuse or circuit breaker protecting that circuit. For example, a 15-A breaker or fuse will protect No. 14 wire; a 20-A breaker or fuse will protect No. 12 wire. As just stated, be sure the wire is properly protected.

After you have brought the end of the Romex to the new receptacle box, a source of power for the receptacle can be a basement location such as a junction box or light fixture. Receptacles located in the kitchen or dining room *must* be on a 20-A circuit. Therefore, the wire *must* be No. 12. The other circuits can be wired with No. 14 wire provided the overcurrent protection (fuse or circuit breaker) is rated at 15 A. Check out existing circuits, and choose the one with the lightest load. In addition to the power sources mentioned above, you can run the cable directly to the distribution

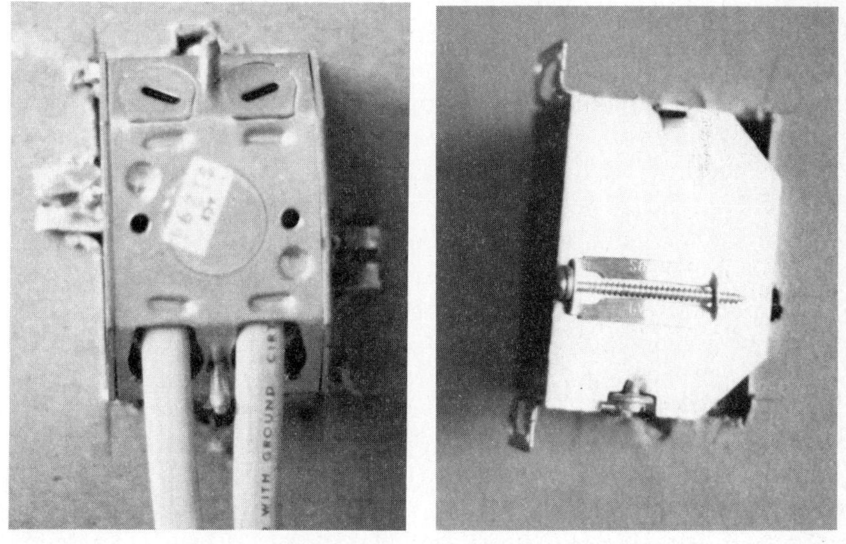

(*a*) (*b*)

FIG. 7-6 (*a*) Wall box showing built-in holding method. This is used mainly in drywall construction. (*b*) Wall box installed in drywall (rear view).

panel. Using a circuit in the panel as a power source, pick a circuit that is lightly loaded.

Originally you checked receptacles and lighting fixtures on each circuit to determine its load. Refer to your record to find a lightly loaded circuit. Just to be safe, remove the fuse or turn off the circuit breaker, and recheck your original findings. If all the existing circuits are loaded to capacity (80 percent of the fuse or breaker rating), you will have to add a new circuit.

> CAUTION: Check the total connected load on the panel so that you do not overload the service entrance cable *and* panel capacity.

Both breaker panels and fuse panels usually provide for extra circuits, the breaker panel having spaces for breakers and the fuse panel having the actual spare fuse holders. If there is space for an additional breaker, record the panel make and model, because not all brands of breakers will fit in all brands of panels.

> CAUTION: Breakers *must* be installed with the main breaker off (for safety).

Most breakers hook under, or spring jaws clamp onto, the power bus bar and then are pushed down into place. Usually installation directions come with each breaker. Many condominiums have the main breaker outdoors near the meter for that unit. The breaker will be under a small hinged cover that has a hasp for a padlock. When you are working on the distribution panel indoors, be sure to turn off the main breaker and lock the cover before working on the panel indoors. Individual houses will, in most cases, have the main breaker indoors at the top of the main service entrance distribution panel, in the basement or utility room. Turn off the main breaker, but remember that the *incoming terminals* of the main breaker are hot (live). Avoid working close to them.

When you have installed the breaker (you do nothing to a panel with fuses at this time because it is ready for connection when the fuse holder is in place), you can run the Romex from the new receptacle to the distribution panel or to the junction box in the basement. Standard wiring practices must be followed. Drill $\frac{5}{8}$- or $\frac{3}{4}$-inch holes through the *center* of the joists in the basement ceiling. Run the cable parallel with the joists, and secure it every $4\frac{1}{2}$ ft.

When connected to a junction box or wall box, the cable must be secured (stapled) within 12 inches of the box if the box has clamps to hold the cable. If there are no clamps, staple within 8 inches. Since the new receptacle is considered old work, the cable need not be stapled as above (this is an exception to the stapling rule). Make the installation neat and workmanlike.

Connection to the end of an existing circuit is easy; just leave 8 inches of slack at the box for easy connection. If the cable enters a ½-inch knockout, you must use a box connector which clamps the cable and is held in the box by a locknut on the clamp. Certain types of boxes have built-in clamps. At the panel you will use a connector. The white (neutral) bar (see Fig. 7-7) may be some distance, inside the panel, from either the new breaker or fuse holder or the knockout you are using to enter the panel; allow plenty of slack, perhaps 24 inches.

There are many types of box connectors. Their main purpose is to hold the cable in place and take the strain off the connected screw terminals in the panel. Distribution panels now have a special equipment grounding bar in addition to the neutral grounding bar (see Figs. 7-8 and 7-9). The equipment bar connects all the *bare* equipment wires (the ground wire in Romex) plus any other special ground wires. This includes the third terminal (green hex screw) on receptacles. Both grounding bars have numerous setscrews, which clamp wires tightly in holes under these screws. The neutral bar will have all the white wires from the cables and the large-size wire from the service entrance cable also. All black wires go to the fuse or breaker terminals.

REPLACEMENT OF CONCEALED WIRING

To replace defective or obsolete concealed wiring (no grounding wire) is a lengthy procedure but not an especially difficult one. Replacement methods will vary depending on specific conditions. Access to an attic, basement, or crawl space will allow you to pull new wiring in place without much trouble. If the dwelling is occupied, work on one circuit at a time. One good reason for replacing wiring is the absence of a ground in the nonmetallic cable already in place.

First remove the fuses or turn the breakers to the off position

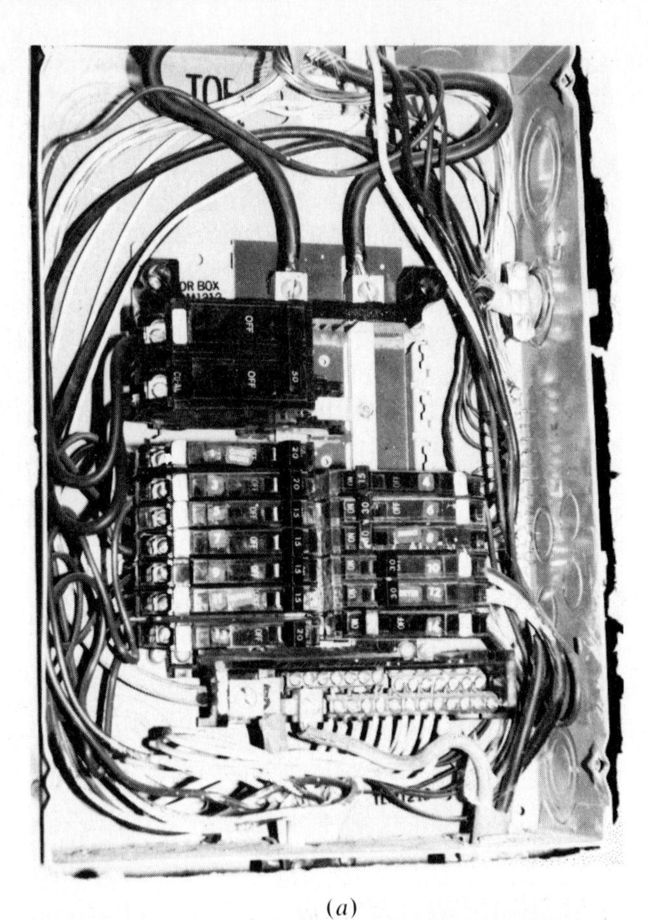

(*a*)

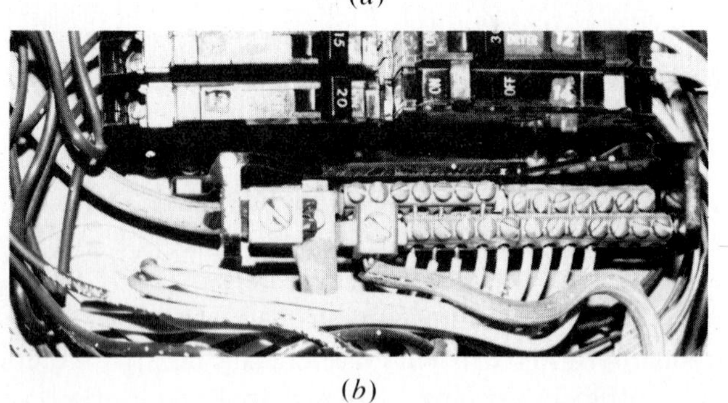

(*b*)

FIG. 7-7 (*a*) Circuit breaker distribution panel. Two large black wires show the feed from the main breaker. The white neutral cable (below breaker area at left) goes to the neutral bar where rows of small screws connect the white grounded wires from the branch circuits to panel. (*b*) Neutral bar, enlarged for detail.

for the main power supply. The breaker may be outdoors at the meter location. Turning off this outdoor breaker makes everything indoors "dead" (no power). If the breaker or fuses are indoors, the wires coming from the meter will be hot (live) (see Figs. 7-10 and 7-11). Notice where these hot terminals are and avoid them; they will be at the top of the cabinet. If you have selected one circuit to replace the wiring, remove the fuse or turn the breaker to the off position for that circuit. Disconnect the black wire from that fuse or breaker, and disconnect the white wire from the neutral bar, where all the white wires are connected. Do not disturb any other wires.

Note: While the panel has no power (except for the two wires to the main fuses or breakers at the top of the panel), check and tighten all connections with wires on them.

All parts of the circuit being rewired in either the attic or basement can be removed by cutting sections and pulling them free. Cut the old cable in short sections so it cannot be used again. Reuse the holes in joists and other framing members for the new cable runs. Where cable goes to receptacles or ceiling fixtures, parts of the cable may have to be abandoned because inside the walls the cable may have been stapled to studs. Remove all the receptacles as you work. Do not disconnect wires; cut them free. The obsolete receptacles will be discarded along with the cable. Remove cable clamps from metal boxes, and cut the cable as close as possible to the knockout openings; push the cable through the opening. In the basement or attic, also cut the cable and push it back into the wall hollow. From the metal box knockout, drop a cord with a weight on the end. Coil a short length of wire solder into a tight coil with a loop on one end, and use it as a weight (mouse). From the basement fish this weight and cord from the old cable hole; tie on the new cable; and pull it up to the receptacle box.

This method will work for removing old cable from the attic down to a wall switch or receptacle, except in this case it is done in reverse order; the cord and mouse are dropped inside the wall from the attic location and fished out through a knockout in the wall box. In rare cases you may have to remove the wall box from the wall. By using a hacksaw blade, you may be able to saw nails holding the box to the stud. Other cases will require ingenious methods either

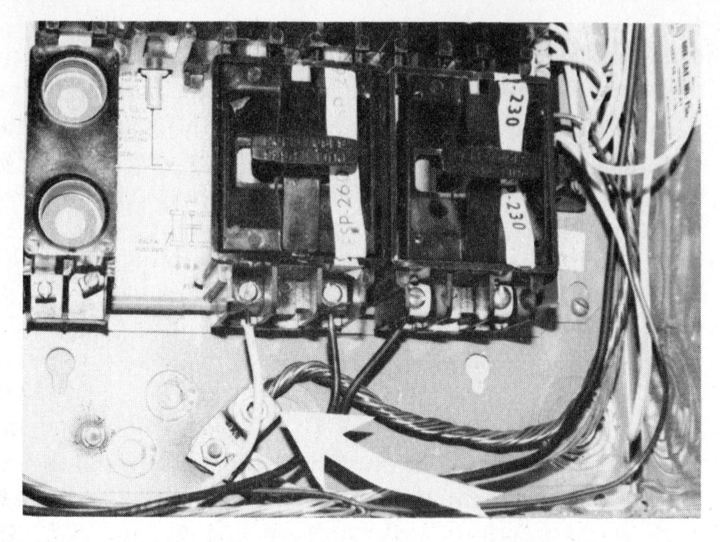

FIG. 7-8 Service cable neutral connection is shown by the arrow. This is the panel in Fig. 5-2.

to "snake" new cables into the box or to remove the box by some other method. Try cutting a larger opening in the bottom of the stud space. Work from the basement or attic, cutting a hole 3 in^2. This opening will allow you to see the box and fish the cord through the knockout into the box. Draw in the cable. Bare about $1\frac{1}{2}$ inches of wire at the cable end, then fold back to form a loop and twist the loop on itself. Thread the cord through the loop and tie securely. Pull it into the box or area wanted, using a helper to guide the cable. The cable is quite stiff and does not pull without someone guiding and shoving at the other end. This type of work is called **old work** because it is installed *after* building construction is completed even if the building is brand new. **New work** signifies electric wiring installed during building construction when the wall finish has not been applied and the ceiling tiles have not been installed. The job is wide open, and wiring can be installed with comparative ease and speed as the building construction progresses.

Whether replacing obsolete wiring or installing new wiring, you must conform with the NEC rules for modern wiring. All wiring must be protected by properly sized overcurrent protection (fuses or breakers) compatible with the wire size and type of insulation. For example, No. 14 wire needs a 15-A breaker or fuse for protection; No. 12 wire needs a 20-A breaker or fuse for protection.

CAUTION: Do *not* use aluminum wire; many problems are associated with aluminum wire, and its use is discouraged. Aluminum wire is only used for service entrance cable. All references to wire in this book apply to copper wire, which is easier to work with.

If the circuit being replaced serves the kitchen and dining areas, you must rewire with No. 12 copper wire to provide a 20-A circuit. Two 20-A circuits are specified by the NEC for this area. Other circuits may be upgraded to 20 A, but you must use No. 12 wire, although you *may* protect for either 15 or 20 A. Many electrical inspectors suggest *all* circuits except the specified 20-A small-appliance circuits be protected at 15 A only.

Abandoned wires or cable must be kept from any contact with the new wiring parts. Push all cut ends out of and away from junction boxes and wall and ceiling boxes. Any circuit can be rewired if necessary. Be sure to purchase and use UL-approved

FIG. 7-9 Main and distribution center (load center). The main breaker is at the top center; neutral bar at far right. Certain breakers are already installed.

FIG. 7-10 Portion of fused distribution service entrance panel. Arrow points to main fuses. Heavy black wires are from the utility meter. These top connections are hot even when the main fuse block is removed.

FIG. 7-11 Removable fuse block with one fuse taken out of the holder. This is the main fuse block; the fuses are 60 A.

devices. Grounding receptacles *must* be used; some toggle switches also have the green grounding screw, and others do not.

CAUTION: Check that there is an equipment grounding wire from the service entrance panel neutral bar to a cold-water pipe (copper or galvanized). A jumper wire must be connected on both sides of the water meter to provide a complete electric path to *ground*. Many improperly wired houses do not have this grounding conductor installed. This is dangerous and should be corrected as soon as possible. The grounding conductor should be No. 6 bare, stranded, copper wire that can be stapled to the wood framing.

Start running the wire at the water pipe, using special ground clamps available at the hardware store or home center. Buy three clamps, since two will be needed at the water meter. Connect the wire to the neutral bar last—*after* you have shut off the power to the entrance panel. Buy enough wire to allow for the meter jumper. String a cord or clothesline along the wire route to estimate the length; measure the cord, and add a percentage to compensate for errors.

Although the absence of a grounding conductor is unlikely, this was the case in the house which I purchased. A test for ground should confirm its installation, for the ground at the utility pole will act as the equipment ground. A visual check of the premises will show the omission. The NEC requires the ground to be made on the premises. There is a small indicating tester on the market which shows defects or omissions in the wiring of receptacles (see Fig. 7-12); it sells for $5–$8 in hardware stores or home centers. It has three prongs and plugs into receptacles; lights indicate errors in wiring. The equipment ground is continuous to all light fixtures and any other non-current-carrying metal parts of the electric system. This ground is connected to the round third opening on receptacles, which accepts the third prong on attachment plugs. As a result, any plug-in appliance and power tool with a three-wire cord and three-prong attachment plug is effectively grounded all the way back to the service entrance panel when it is plugged into this grounding outlet.

Note: Double-insulated power tools are excluded because they only require two-prong plugs.

Figure 7-13 shows a continuous path of the grounding conductor from the hand power tool all the way back to the neutral bar in the service entrance panel. From this bar a **grounding conductor** extends to a metal cold-water pipe (usually copper) which in turn is buried in the earth outside the building. If there is a water meter in the water line, it *must* be bypassed with a permanent copper wire jumper. Two ground clamps, one on each side of the meter, pro-

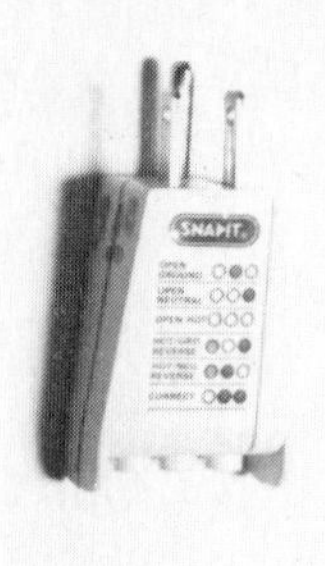

FIG. 7-12 This tester indicates incorrect wiring of receptacles obtained by just plugging into the outlet. Indicating lights denote what is wrong.

vide a path for fault currents to travel to ground (earth) even though the water meter has been removed.

To develop an effective grounding system, all metal parts used in the building construction must be bonded together to form an integral ground. This will include the metal water piping from the city water supply, the metal part of the building structure, any gas piping (check with the inspector and the gas utility before connection to the gas piping), and all other metal piping, as well as metal air ducts. A driven rod or pipe is also required. The pipe is required to be galvanized $\frac{3}{4}$ inch. Steel rods must be $\frac{5}{8}$ inch in diameter and nonferrous rods $\frac{1}{2}$ inch in diameter. All pipes or rods must be driven far enough to have 8 ft in contact with the earth.

GROUND FAULT CIRCUIT INTERRUPTER

The GFCI was developed in the last decade to protect people from shock incurred while using potentially defective portable appliances, such as hand mixers, hair dryers, and portable power tools. Even if these power tools have three wires and three-prong plugs, additional protection is provided by the GFCI. This device acts much faster than either breakers or fuses and responds to very small current flows. While GFCIs are more expensive than other overcurrent devices, their value in saving lives makes the cost worthwhile.

GFICs are made in three types: (1) **wall receptacle,** (2) **circuit breaker,** and (3) **portable.** All types have a test button which must be pushed once a month to test operation when a fault occurs. These devices will operate almost instantaneously to protect people and property. The receptacle type is commonly used in bath-

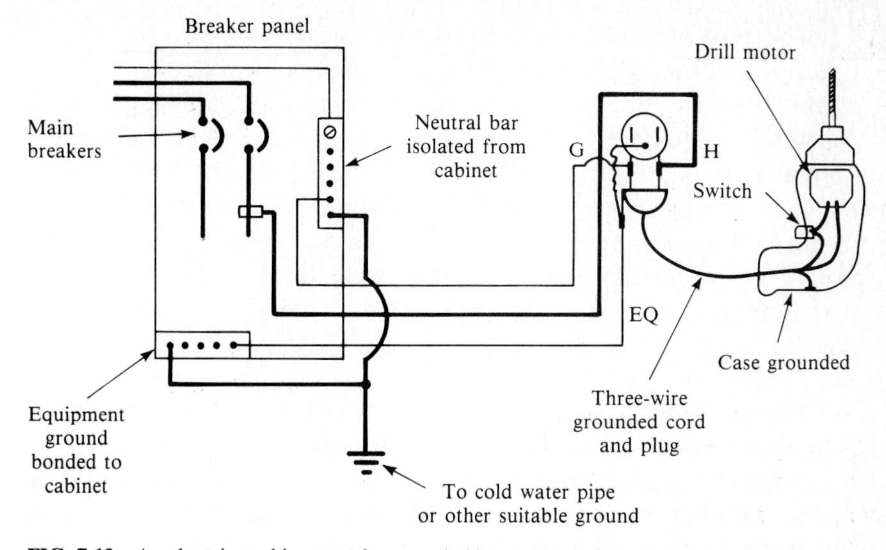

FIG. 7-13 An electric tool is properly grounded by connecting equipment ground wire with the actual ground at the service equipment.

rooms and is supplied in *receptacle-only* protection or *receptacle and downstream* protection. *Downstream* signifies wiring and devices from that point and beyond *away* from the panelboard. The circuit breaker type protects the complete circuit connected to it. The portable type is used on construction sites to protect the portable tools used by the workers. Both the receptacle type and the breaker type replace their respective types directly. The receptacle type is bulky and may need a raised cover plate to fit into the wall box properly.

The NEC requires that GFCI protection be provided for the following receptacles:

- In bathrooms.
- In garages. Protection is not needed for receptacles that are not accessible or those used for freezers, washers, or dryers.
- Outdoors where there is access to a person standing on the ground.
- Construction sites.
- Other locations may be protected at the option of the homeowner, such as the kitchen receptacles.
- Swimming pools. Pools are subject to very stringent requirements. Contact your local electrical inspector for guidance.

The NEC recommends that GFCIs be installed to protect all outdoor receptacles where appliances can be plugged in by persons standing on the ground or concrete (concrete is considered to be "ground") or where contact with any metal which is grounded can be made. I advise protecting the kitchen receptacles with a GFCI also. As noted before, the NEC only makes recommendations, so requirements will vary with the governing body having jurisdiction. GFCIs should be installed where the danger of shock exists. The amperage rating of the GFCI must correspond with the wire size being protected. A No. 14 wire needs a 15-A GFCI, and No. 12 wire needs a 20-A GFCI.

Turn off the power when you install either GFCI. The breaker type requires that the entire panel be dead. Turn off the main breaker to install the breaker type. Be careful when you remove the panel front. Do not drag it into the panel wiring.

SYSTEM CAPACITY

Although only the present wiring is to be replaced, it is still wise to check whether the service entrance equipment and wiring from the service drop are sufficient to handle the wiring already installed. Note that additional wiring may have been added by previous owners. An example of a load calculation for a dwelling follows:

The dwelling has a floor plan area of 1000 ft^2, not counting the basement, attic, or porches. It has a 12-kW range and a 2.5-kW water heater. The clothes dryer is gas-operated.

1000 ft^2 at 3 W/ft^2		3.0 kW
Two 20-A appliance circuits, 1500 W each		4.0 kW
Laundry circuit		1.5 kW
Range		8.0 kW
Water heater		2.5 kW
Total		19.0 kW
First 10 kW at 100%	10.0 kW	
Remainder at 40% (9 kW × 0.4)	3.6 kW	

The calculated load for a service size of 13.6 kW is 13,600 W. Divide 13,600 W by 240 V to get 56.7 A. If this dwelling has 60-A service entrance equipment, it may continue to be used. Be aware that any additional new appliances or other equipment may overload the service equipment. You may want to consider installing new service at this time. Note that this calculated load will probably need to be increased to meet the demands of today's living standards. The NEC recommends only *minimum* requirements. Design the system for proper capacity of present-day needs.

Wiring Methods and Special Techniques

Any mechanically inclined person can repair, remodel, or add to the wiring in a home. Sometimes, however, the advice and explanations of an experienced electrician will save you time and money and ensure a top-quality installation. Any trade or profession has shortcuts or more efficient methods of executing a certain procedure.

Consider the following example. The project is to install a ceiling fixture in the center of the ceiling of a room 14 ft 6 in × 15 ft 1 in (see Fig. 8-1). Locate a thin stick of wood 6 or 7 ft long. Set a stepladder near the center of the room. Climb the ladder high enough to reach the ceiling; shove the stick against one wall where it meets the ceiling. Mark a point on the ceiling where the board ends. Shove the stick in the opposite direction (don't change ends) as you did before, and make a mark. Now do the same crosswise of the room, making two marks as before. You now have four marks, two in one direction and two crosswise. With a tape measure or rule, measure the space in each direction and divide each in half.

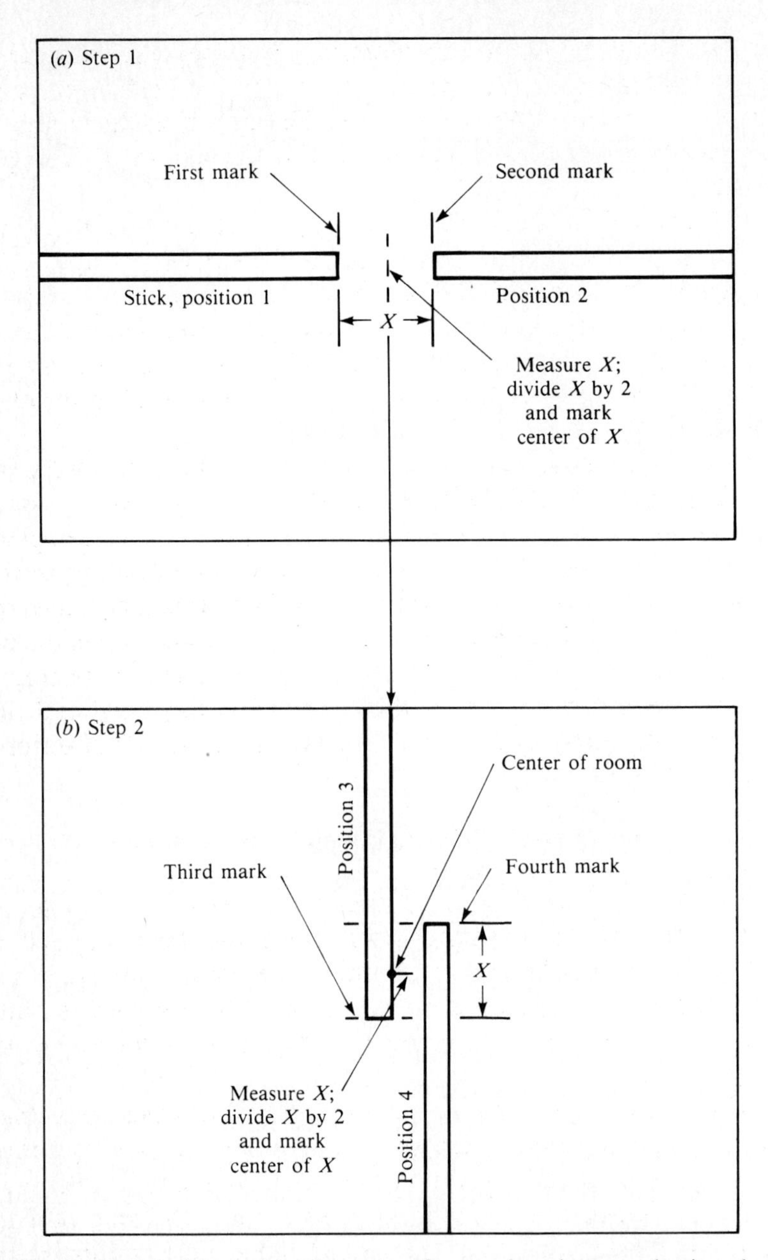

FIG. 8-1 Method of determining the center of a room by using a wood strip and rule.

This will be the center in each direction. You may have to slide the centers so they coincide. Be sure to identify your final marks so that you do not use the wrong ones. The directions are long, but the procedure takes about 5 minutes. Try this on a small ceiling for practice.

In an actual installation, there is a 50–50 chance that the center will fall on a joist. There are ceiling boxes only ½ inch deep which will usually fit flush with the plaster or drywall. If not, chip away some of the joist. Use screws to mount the box directly on the joist. Be sure to allow space for the electric cables to enter the box beside the joist. Figure 8-2 shows various items used in installing wall boxes.

If the box location lies *between* the joists, support for the box must be provided. The best is a steel support bar, of adjustable length so as to span the joist space. When the bar is fitted in place, the bar ends are nailed to the joists. The box can slide along the bar to align with the ceiling opening. Heavy fixtures *must* have this support. Another box has a U-shaped bracket which, when the box is pushed into the wall opening, expands wider than the opening. A screw in the center of the back of the box is tightened and thus pulls the box snugly into the opening (see Fig. 8-3). This action is similar to that of a toggle bolt.

Note: This support is for lightweight fixtures only.

Also when you are hanging fixtures from plastic boxes, do not utilize the tapped holes in the box sides, because these may pull out if the box must support a heavy fixture. These holes are used when the slotted bar furnished with fixtures is used across the box width. Either use a metal box, or screw an adapter onto the stud in the bottom of the box.

Wall boxes have a similar spring-out bracket. Other boxes have built-in brackets which, when screws on either side are tightened, pull wings that spread out like a Molly screw anchor. A separate bracket is shaped like the Greek letter π (pi). The long part fits behind the plaster, and two short legs bend around and inside the box, securing it to the wall.

Where studs or joists are exposed (new work), boxes have angle brackets that are to be nailed directly to the stud. Plastic boxes have ears on the top and bottom outside for holding nails to

FIG. 8-2 Wiring items. Top row: Bracket for plastic wall box, BX clamp, Romex clamp, Romex clamp (used with wall boxes). Bottom row. Wall box with integral support, support pulled up (like Molly bolt), Madison wall box support.

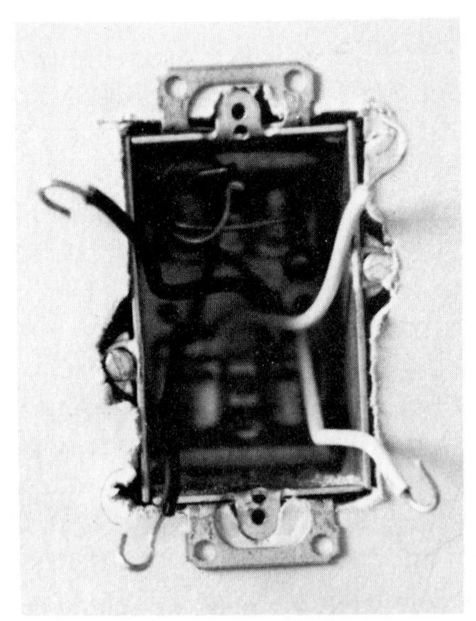

FIG. 8-3 Metal wall box installed with Molly-type pull-in anchors. Anchors are integral with the box.

be pounded into the side of the stud. The box must be positioned to end up flush with the wall surface, drywall, or other material. Determine the wall material thickness before mounting the box.

Sometimes it is necessary to remove a box from its location in a finished wall. This could be difficult if the box has a welded bracket nailed to the front of the stud. If the other side of the wall is open, as in a basement or attic, it will be easier. Other boxes may be nailed, as mentioned above, to the side of the stud. With a screwdriver, try to pry the box away from the stud far enough to squeeze in a bare hacksaw blade and saw through the two nails, freeing the box. Plastic boxes usually have this type of support also. To reattach the box, drive two round-head screws through the side of the box into the stud. Some boxes have holes; others do not. Any holes used must be close to the box front because a screwdriver must be able to reach these screws. When the box is back in place, any openings around it must be sealed with patching plaster, as required by the NEC.

INSTALLING AN ADDITIONAL RECEPTACLE

The method of removing a wall box is sometimes used in extending a circuit to provide an additional wall receptacle. When deciding on the location of the receptacle, you should consider the route you must follow in extending the cable to it. Sometimes a nearby location requires that the cable take a longer route owing to obstacles along the route. The wood framing must be slotted to accept the cable and the cable must be protected by $\frac{1}{16}$-inch metal under the trim. In extreme cases, the trim around a door frame must be removed and a groove cut for the cable to install a receptacle beyond the doorway. Houses on concrete slabs and with eaves all around require this type of routing (see Fig. 8-4).

When a receptacle is desired on the opposite side of a wall from an existing receptacle (back to back), the job could be easy. If the wall space is free, cut the opening. Drill a small hole, and probe with a wire to check for obstacles first; then cut. You can reach in the hole and usually twist out a knockout in the present box, using a stubby screwdriver. In this case the box need not be removed.

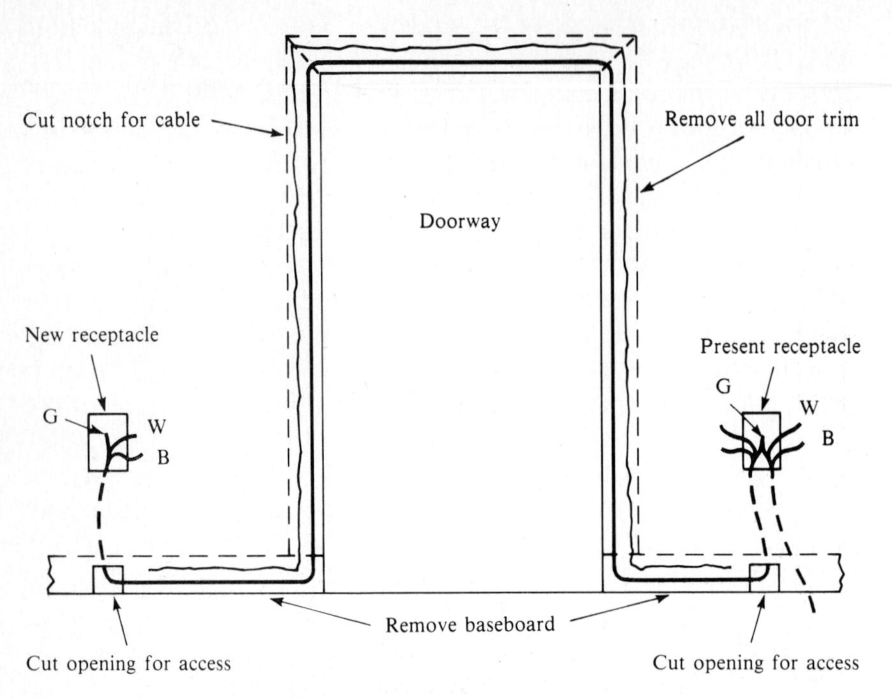

FIG. 8-4 Method of routing cable around a doorway.

> CAUTION: Do not install the box exactly opposite the present box because both boxes will exceed the wall thickness, and you will have trouble. Offset the new box as far as possible in either direction (see Fig. 8-5).

The bending of the cable will also be difficult if the boxes are too close together. Turn off the power to the present receptacle; disconnect the receptacle. Reach in through the new opening, and remove the knockout in the present box. Thread cable into the box; allow 8 inches for making connections. Cut a section of cable and allow for plenty of length. If you removed a bottom knockout, do the same with the new box so that the cable will loop *downward* when you install the new box. Top knockouts will allow the cable to loop *upward*. Depending on the anchoring method, fasten the box in place. Connect the new receptacle first, white wire to silver screw, black wire to brass screw, and bare ground wire to green screw.

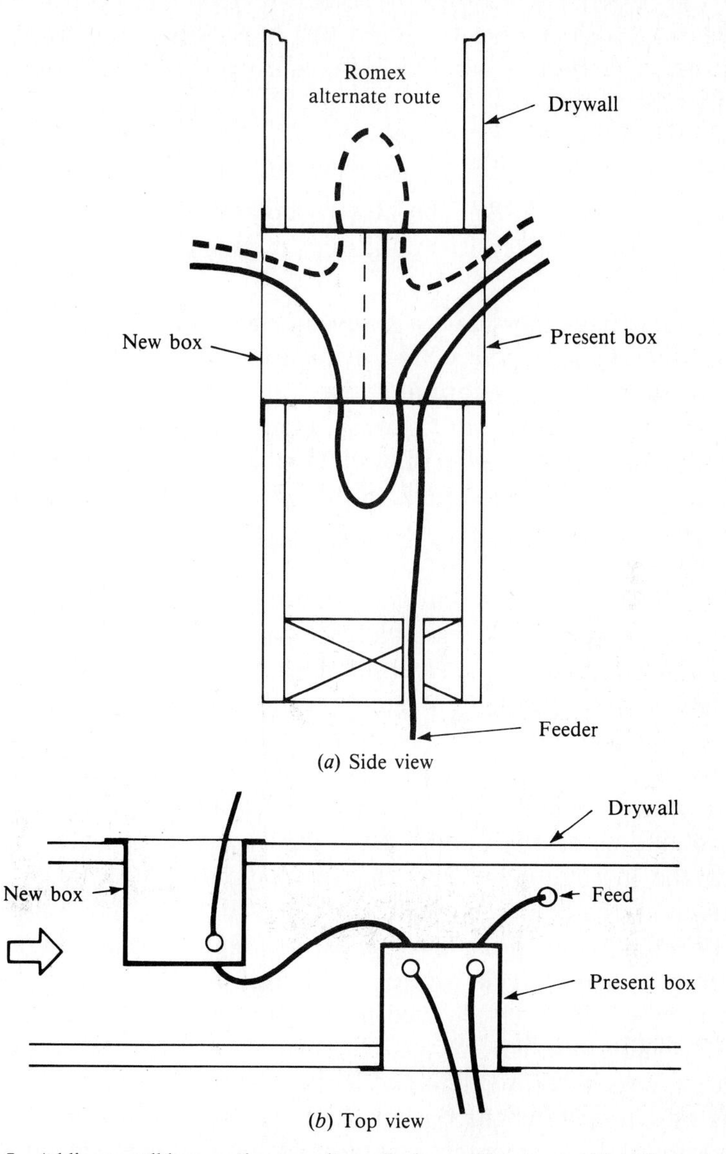

FIG. 8-5 Adding a wall box on the opposite wall of a partition to avoid interference of the boxes.

Note: If the present receptacle has only two slots, refer to the section on grounding before attempting this project.

Now reinstall the present receptacle. This receptacle will have two screws on each side—two silver, two brass—and one green on the bottom. The bare grounding wire will need a pigtail and a wire connector to connect:

1. The present receptacle
2. The new receptacle
3. The box, if it is metal

The metal box may have a screw in the back of the box for the grounding wire, or it may have just a tapped hole. This box *must* be connected to the bare ground wire.

Access from the attic is easy since the wiring is usually exposed, making the tracing of circuits simple. Basement wiring may be exposed in some areas and not in others. The average home will have six to eight circuits. Two small-appliance circuits to serve the kitchen area are mandatory, another NEC requirement. These are rated at 20 A and are wired with No. 12 copper wire. All other circuits are general lighting circuits rated at 15 A and wired with No. 14 copper wire. The small-appliance circuits can have as many as ten receptacles each, while the other circuits should have a maximum of eight.

When you are picking a circuit to extend, select one having the fewest receptacles or the least used receptacles. This will result in equal loading of all circuits. The simplest extension of a circuit is from the last receptacle on the run. If you tap in at a receptacle midway in the run, you may crowd the wall box. In this case there will be nine wires in the box—three coming in, three going on, and three new ones (black, white, and bare). The NEC now specifies the number of wires allowed in a junction box of each size (see NEC, Table 370-6a). Since all current-carrying wires generate heat, excessive crowding may damage the insulation on the wires. It is best to extend the circuit from the last receptacle and backtrack to the new receptacle location. Most main panels have spare circuits (fuse type) or spare places for an additional new circuit breaker. Buy and install a new breaker and be sure to use the correct size breaker for the wire size you are using—15-A breaker for No. 14 wire and 20-A breaker for No. 12 wire.

In this case the receptacle can have its own circuit starting from the main panel; other receptacles can also be added. When you buy a new breaker, the package will have directions for installation.

> ***Note:*** The new breaker *must be listed* for use in *your* panel. The panel model number and name will be listed if the breaker will fit.

> **CAUTION:** Be *sure* to turn off the main breaker before you try to install the new branch circuit breaker.

If in the panel there is a breaker or fuse feeding a lightly loaded circuit, the connection may be made to that circuit by mounting a $4\frac{11}{16}$-inch square box beside the panel (see Fig. 8-6). Use an offset nipple to connect the box to the panel enclosure. Knockouts are seldom in line, so the offset nipple is necessary. Disconnect the Romex cable from the breaker or fuse holder terminal, the grounding bar, and the neutral bar.

> **CAUTION:** Be sure you have removed the main fuse or turned the breaker to the off position before starting this work.

This cable is then brought into the new junction box (use the appropriate clamps) along with the new cable to feed the new receptacle. Thread the three wires from a short length of cable, not the cable itself, because these wires will go through the offset nipple. Be sure these wires are not too tight and are long enough to reach to their respective terminals with easy bends. Attach black to the fuse or breaker terminal, white to the neutral bar, and bare to the grounding bar. Returning to the new junction box, connect the wires, color to color, with wire connectors. I like to tape the skirt of each wire connector with electrical tape as a precaution against short circuits. The new box is already grounded through the offset nipple.

Sometimes a new circuit has to be run from the basement to the second floor to supply a receptacle needed there. To do this, you will have to be resourceful to find a route from the basement to the second floor. Stay away from heating ducts and hot pipes. The new plastic insulation will melt and cause a short circuit. If the new location is in a partition which can be reached from an unfinished basement area, drill up into the hollow space. Find a length of stiff

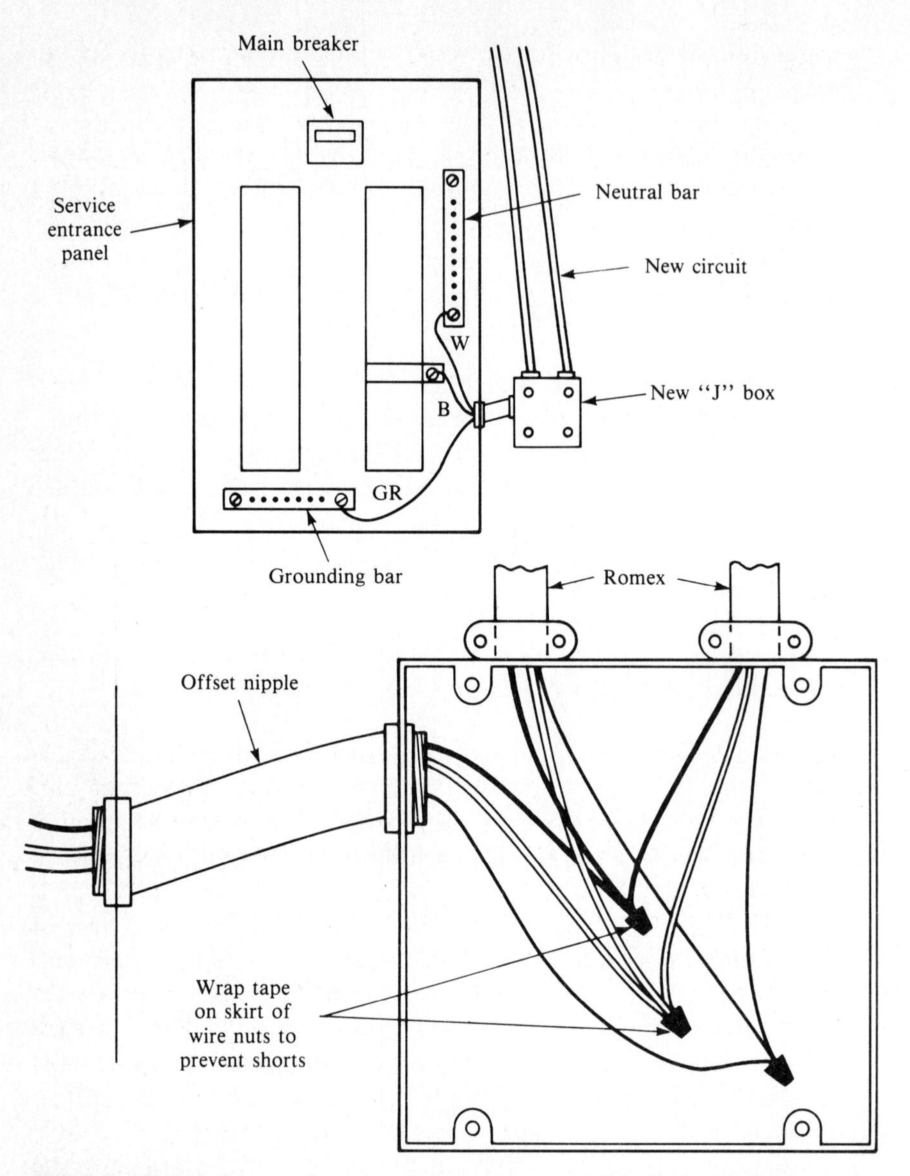

FIG. 8-6 Method of connecting a new receptacle to an existing circuit, starting from the main distribution panel.

wire about 10 ft long (No. 11 clothesline wire), form a loop on one end as a handle, and probe up inside the wall. If you go up 8 ft, the wall is free of obstacles. By very careful measurement determine where on the second floor you are in relation to this hollow wall space. When you are *absolutely* certain everything is lined up (some second-floor partitions are offset from those downstairs), cut the wall box opening. As shown on Fig. 8-7, you may be able to use an 18-inch bit extension with a $\frac{3}{4}$-inch spade bit. The drilling angle must be almost vertical, or else you might come out in the downstairs room opposite. If you cannot go this route, take off the baseboard and cut out until you can drill down through from this point. *Think out* your procedures, for you will have to repair your mistakes. All this work goes much better if you have a helper.

WORKING WITH CABLE

The two types of cable that the homeowner will use are nonmetallic sheathed cable (Romex) and Armored Bushed Cable (BX). Both are used for the same purposes and are identical except for the outer covering.

Nonmetallic Sheathed Cable

Current cable construction is of two types: **NM**, in which the wires are enclosed in an outer sheath along with fiber wrapping around each wire, and **NMC**, in which the two insulated wires and the bare grounding wire are embedded in a solid plastic sheath. This may be used in damp but not wet places. Either type is easy to work with. Cut it with 8-inch diagonal cutters or side cutters. To strip the sheath, a cutter is available at hardware stores at about $1.50. This cuts down one side of the sheath, allowing it to be peeled back and cut off (see Fig. 8-8*a*). Romex is sold in 25–250-ft lengths. Prices: 14-2 with ground, 10¢–12¢ per foot; 12-2 with ground, 12¢–15¢ per foot. Watch for sales specials.

Cable run in new construction is threaded through holes bored in framing members. The NEC requires that the holes be in the center of the wood members (not close to one edge). This is to prevent nails being driven into the cable (see Fig. 8-8*b*). Cable may also be run in notches cut in the edge of the framing members, but then metal plates $\frac{1}{16}$ inch thick must be nailed over the cable in the notch. These *cable protectors* are available at most hardware stores.

Cable may be run on the side of any framing member; it must be stapled every $4\frac{1}{2}$ ft and within 12 inches of a metal box or within

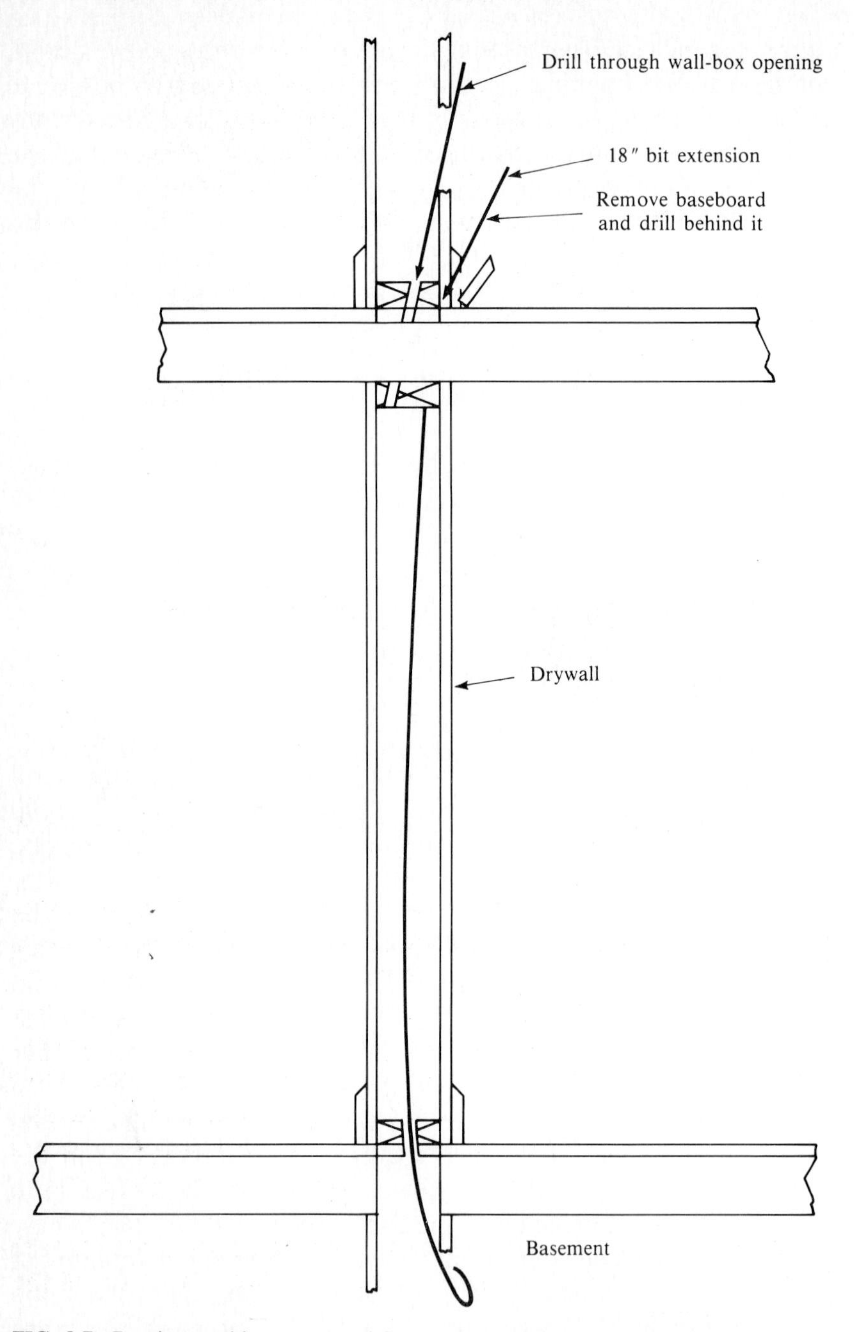

FIG. 8-7 Routing a cable to a second floor. Note drilling method used to gain access to second-floor partition.

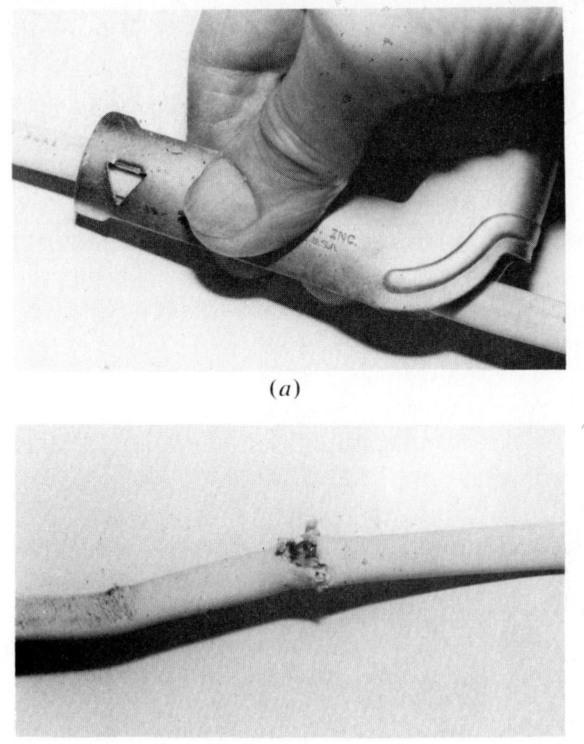

(*a*)

(*b*)

FIG. 8-8 (*a*) Slitter to strip cable sheath. (*b*) Typical damage to nonmetallic sheathed cable.

8 inches of any plastic box not having cable clamps. Use metal staples (don't pound in too tightly, or you risk deforming the cable) or the new plastic style; this is a plastic bar with a small nail through each end. Although the plastic staple is better, do not pound the nails too tightly. Some electrical inspectors recommend that the cable be able to be moved slightly through either style staple. Running boards can be used to carry cable across framing members. These are used in attics near the access opening. In this case the boards should form a trough to fully protect the cable.

> ***Note:*** Do not bend cable sharply, for this is prohibited by the NEC. The bend radius must not be smaller than 5 times the cable diameter.

All metal wall and ceiling boxes either have cable clamps integral to the box or use an external separate connector. An assortment of connectors is shown in Fig. 8-9. The connections to a duplex receptacle are shown in Fig. 8-10. The green hex grounding screw is shown on a receptacle in Fig. 8-11. In most boxes, knockouts are now "pry-outs"—they have a slot to accept a screwdriver to pry or twist the slug out. I once had a shipment of $4\frac{11}{16}$-inch square boxes which had poorly punched knockouts (not punched deeply enough) and I had to use a heavy hammer and chisel to "knock out" the knockouts. Many times the box side would deform. I finally used a block of wood to support the box side when hammering on the knockout.

> ***Note:*** All metal boxes must be made a part of the equipment ground by inserting a ground screw in the special tapped hole in the box back.

Metal parts of devices (non-current-carrying) must be grounded by using the green hex screw connected to the bare ground wire in the cable. Plastic boxes may be used with nonmetallic cable; some boxes have clamps, others do not. As with any device, always ground the device to the bare ground wire in the cable.

> CAUTION: *Never* use cable *without* the ground wire; this is still found in some hardware stores.

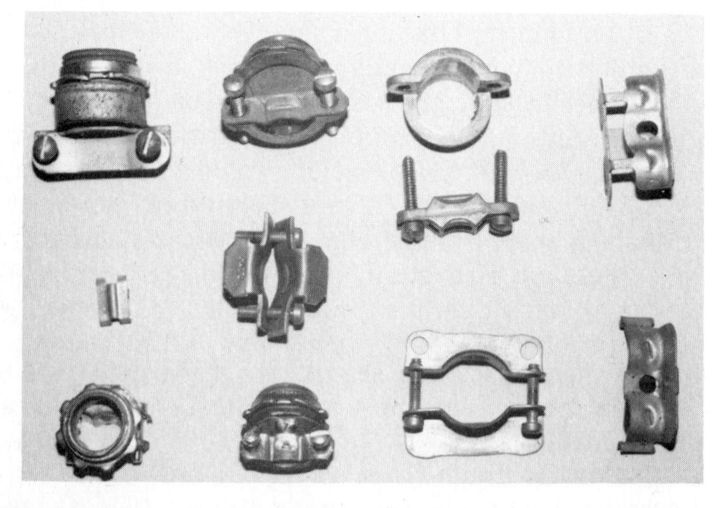

FIG. 8-9 An assortment of cable connectors. All are for Romex cable except the one at the upper right-hand corner, which is for BX.

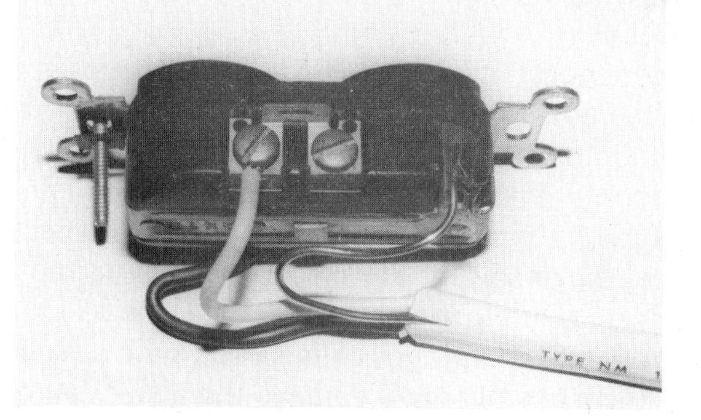

FIG. 8-10 Connections to a grounded receptacle.

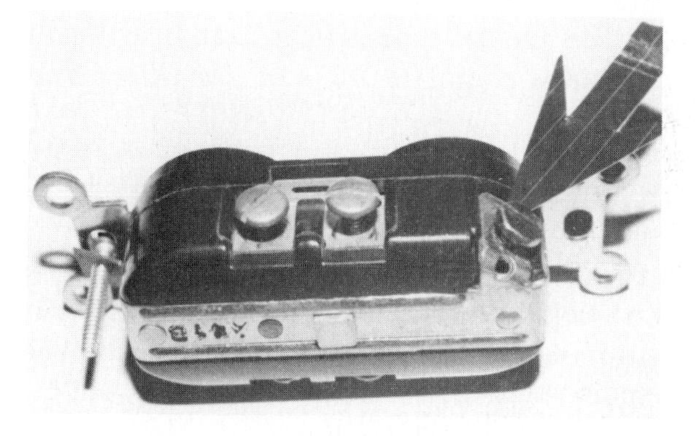

FIG. 8-11 Duplex receptacle; arrow shows green grounding screw.

Newer devices have a specially designed yoke mounting screw which will provide ground continuity from the box to the yoke even though the yoke does not contact the box because it is recessed in the plaster. On exposed wiring where handy boxes are used, the device yoke will contact the box ears and be effectively grounded. In all wiring, always leave enough wire extending from any junction box or device box to make the proper connections; leave at least 8 inches. In this case longer is better because 1 inch too short may cause you to have to run a new length of cable. At all times, consideration must be given to protecting the cable from physical damage. Even if exposed cable goes through a floor, it must be protected by tubing a distance of 2 ft above the floor level.

Armored Bushed Cable

This cable, commonly known as BX, is required in many municipalities because of local code restrictions. BX is more resistant to damage than Romex because of its construction, but BX is installed in much the same way. Boxes must be metal and have the special BX-type clamps or use BX-type box connectors. Both types have openings so that the inspector can see the anti-short-circuit bushings required by the NEC. The metal sheath of BX either is cut with a special tool or is sawed with a hacksaw. This leaves a sharp edge which may damage the insulation on the individual wires. To protect the wires, a special fiber or plastic bushing is inserted between the wires and the armor forming a rounded edge. Since the bushing is red, it is easily seen through the inspection openings. The ground tape is bent back over the armor to be clamped in the box clamp.

INSTALLING ARMORED BUSHED CABLE*

The Armored Bushed Cable shown in Fig. 8-12 is also known as BX, is easy to work with, and lends itself to most wiring jobs.

Locate and mount boxes as shown in Fig. 8-13. The NEC requires a grounding-type convenience receptacle outlet wherever cord-and-plug connected equipment is used. And a receptacle must be installed for every 12 ft (or less) of wall space around a room in a dwelling. In addition, boxes are needed for ceiling outlets. Then switch boxes must be located and mounted. Follow a wiring plan to determine the size and location of boxes, size of conductors, and number of branch circuits.

> ***Note:*** Metal outlet boxes must be used with Type AC cable. (See the NEC, Sec. 370-3.)

Each box is effectively grounded when the cable armor is clamped or secured to the box by the internal box clamp or by a connector attached to a box knockout. Box sizes are determined by the num-

* Text adapted from AFC/A Nortek Co., New Bedford, Massachusetts.

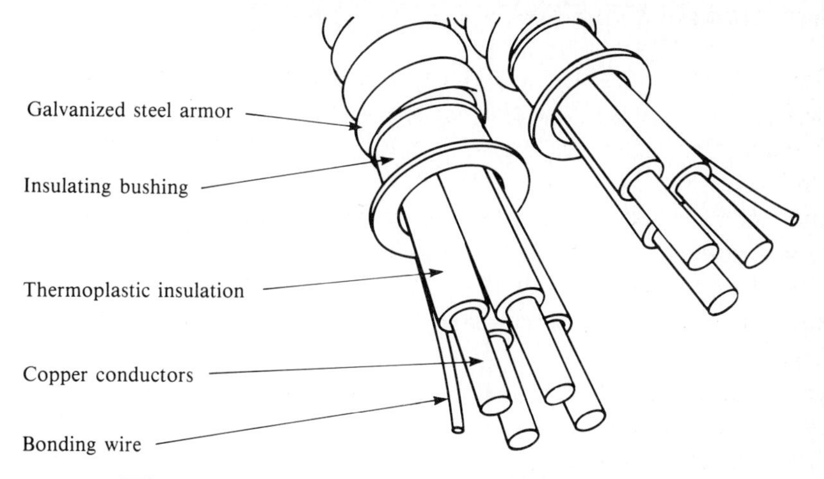

FIG. 8-12 Armored cable. (*Courtesy of AFC/A Nortek Co.*)

ber of conductors, clamps, hickeys, devices, etc. according to the provisions of the NEC, Sec. 370-6a,b and Tables 370-6a and 370-6b.

In drywall construction, be sure to mount the boxes flush with the wall finish on combustible materials and recessed not more than $\frac{1}{4}$ inch for noncombustible materials. To center the ceiling outlet box, use a measuring device (see Fig. 8-1). Ceiling boxes require the use of some type of support bracket, preferably one which allows free access to all the knockouts in the box.

With all boxes installed, simply find the most direct route to conserve both cable and effort. If needed, bore holes where cable crosses joists or studs (metal studs are prepunched). To bore holes in wood, use an extension bit for less slant and easier pulling than a regular-length bit. Or, use an electric drill or a joist boring machine for a power assist. Keep holes as close to the center of joists or studs as possible, with the edge of every hole at least $1\frac{1}{2}$ inches from the edge of the stud or protected by a steel plate. *Remember*: The holes must be larger than the cable diameter so that the Armored Bushed Cable can be pulled through easily (NEC, Sec. 300-4.). Also see Figs. 8-14 and 8-15.

Next, draw the end of the Armored Bushed Cable from the center of the coil in a counterclockwise direction to prevent kinking. Thread cable through the holes in studs. All cable runs must be

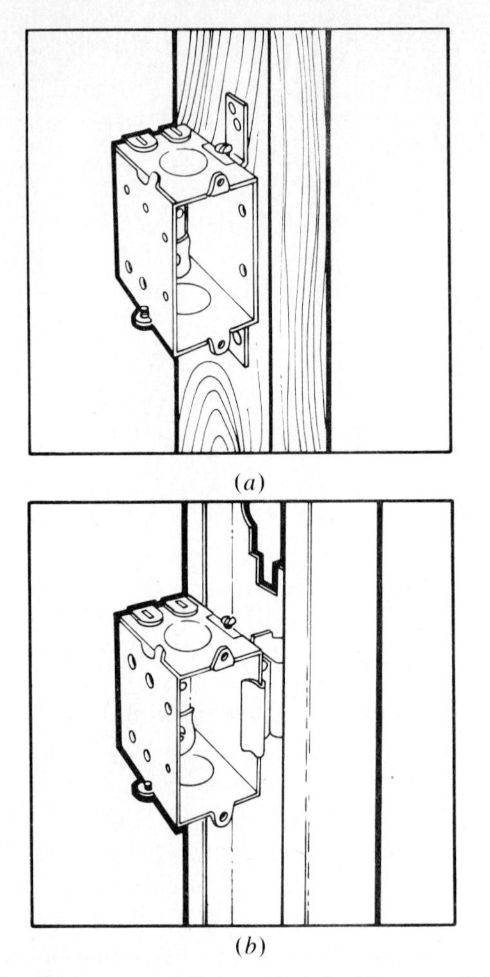

FIG. 8-13 Metal wall boxes mounted on metal studs. (*Courtesy of AFC/A Nortek Co.*)

continuous from outlet to outlet. The cable must be supported every $4\frac{1}{2}$ ft or less, unless it is routed through bored or punched holes, and within 12 inches of every box, unless the cable is fished (NEC, Sec. 333-7). Staples and straps can be used to secure cable to surface to be wired over. When you install cable, bends should be made to prevent cable damage. The radius of the curve of the inner edge of any bend should not be less than 5 times the cable diameter (NEC, Sec. 333-8). When cable reaches a box, cut sufficient cable length that there will be at least 6 inches of free conductor left at each outlet and switch point for connections or splices. A hacksaw can be used to cut the armor for stripping the armor end,

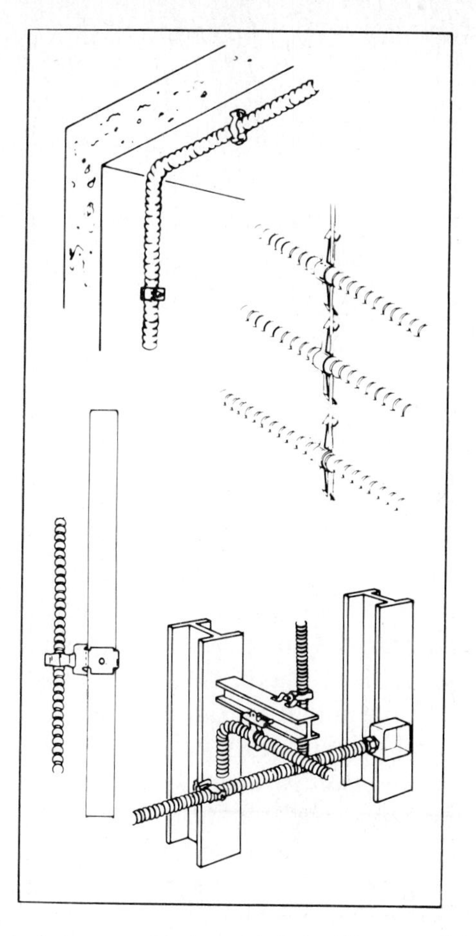

FIG. 8-14 Attaching techniques—cable to metal studs. (*Courtesy of AFC/A Nortek Co.*)

but care must be exercised *not* to cut into the insulation on the conductors. Cut through the armor only, and slip the short piece of armor off the end.

Armored Bushed Cable enters a box through a knockout. Remove the knockout with a tap of a hammer or screwdriver and a twist of pliers. Where connectors are used to secure cable to boxes, ensure a proper bond by firmly tightening the connectors to both the box and the armored cable. Where boxes have internal cable clamps, look for the designation "A" on the clamps, which verifies the suitability for use with armored cable (see Figs. 8-16 and 8-17).

The purpose of the bonding conductor is to ensure that the

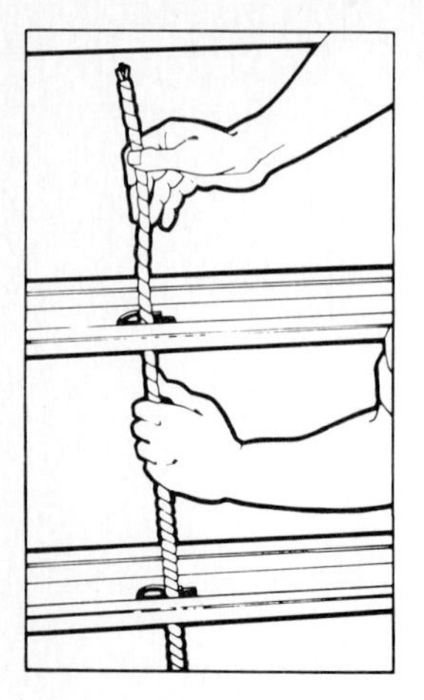

FIG. 8-15 Armored cable being threaded through metal studs. (*Courtesy of AFC/A Nortek Co.*)

ohmic resistance of the armor is low and reasonably uniform throughout the entire length of the cable; that it will remain practically constant if the convolutions of the armor might loosen as a result of flexing during handling; and that it supplements the effectiveness of the spiral armor as a means of equipment grounding.

At each cable termination, simply insert the red anti-short-circuit bushing, and bend back the exposed length of bond wire for either internal clamp termination of the cable or termination with a separate armored cable box connector. Feed the conductors into the box through the clamp or connector mounted in a knockout. Clamp armor securely with box clamp or separate box connector. Be sure the anti-short-circuit bushing is plainly visible in the armored cable connector for easy inspection (NEC, Sec. 333-9).

> ***Note:*** Some boxes have built-in clamps, but external connectors are sometimes preferred because they leave more room in the boxes.

The NEC specifies that a box with internal clamps be limited to one conductor *less* than a box of the same size without the internal clamp (NEC, Sec. 370-6a). Fasten cable to studs at appropriate intervals. Figure 8-18 shows an external connector with the inspection slot to show the anti-short-circuit bushing. Figure 8-19 shows typical examples of armored cable connectors.

This completes the roughing in of the job which is now due for inspection. Most state and/or local regulations require that an elec-

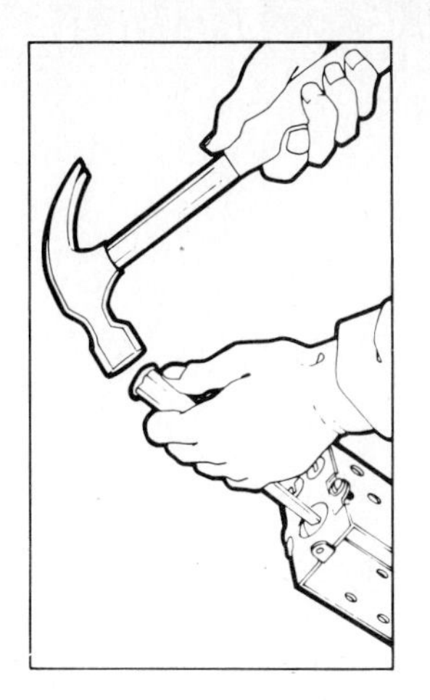

FIG. 8-16 Removing a knockout with a chisel and hammer. (*Courtesy of AFC/A Nortek Co.*)

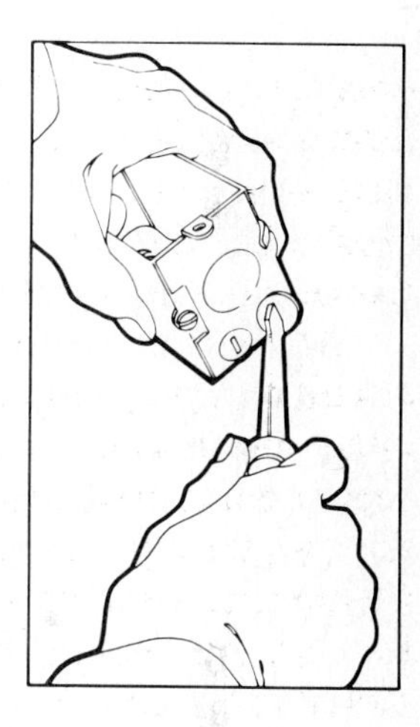

FIG. 8-17 Removing a pry-out knockout. (*Courtesy of AFC/A Nortek Co.*)

trical inspector or approved authority approve the job with a signed, dated identification and approval tag.

After interior walls are finished on the job site, proceed to hang fixtures, mount and connect switches and receptacles, and put outlet covers in place (see Fig. 8-20). Connecting the conductors to fixtures is easiest with wire connectors (wire nuts). Remove just enough insulation for the connector to cover the bare wires. Place the connector over the wire ends, and turn as you would a nut, until tight. Check that all wires are locked in and no bare wires are exposed.

Boxes used with armored cable are shown in Fig. 8-21 with the

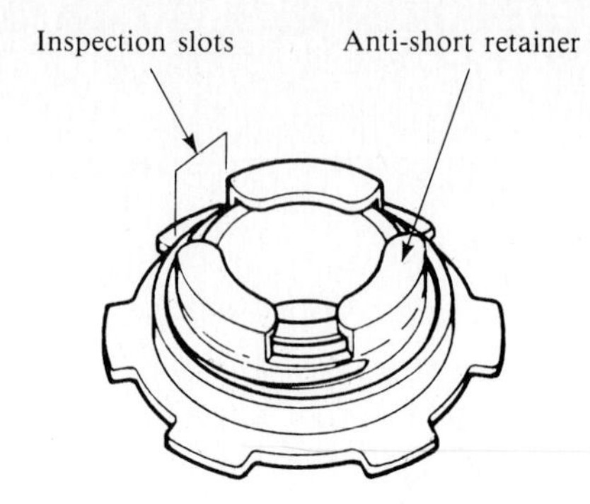

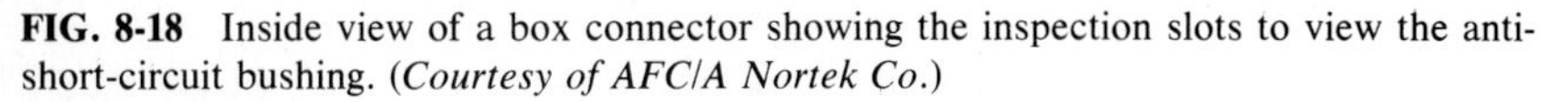

FIG. 8-18 Inside view of a box connector showing the inspection slots to view the anti-short-circuit bushing. (*Courtesy of AFC/A Nortek Co.*)

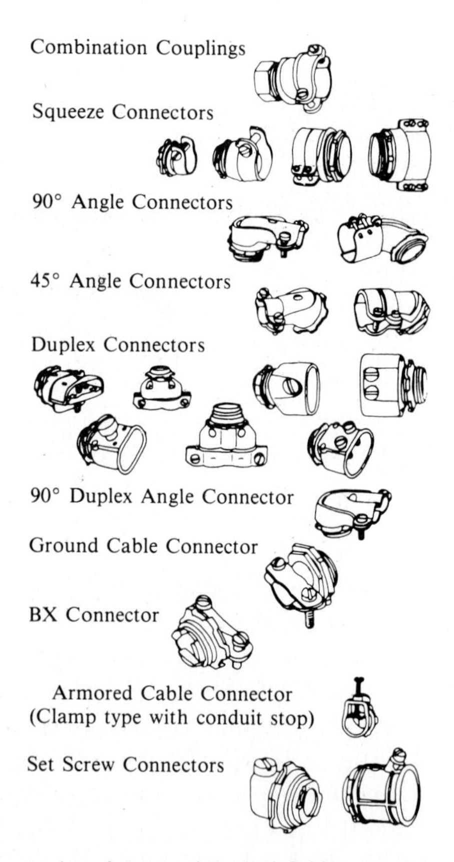

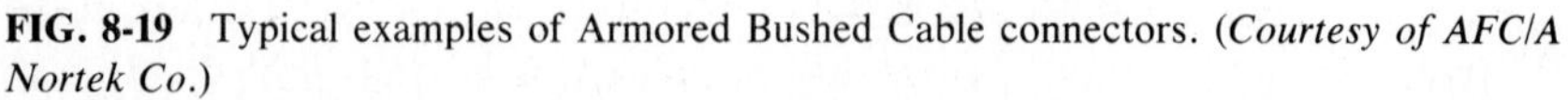

FIG. 8-19 Typical examples of Armored Bushed Cable connectors. (*Courtesy of AFC/A Nortek Co.*)

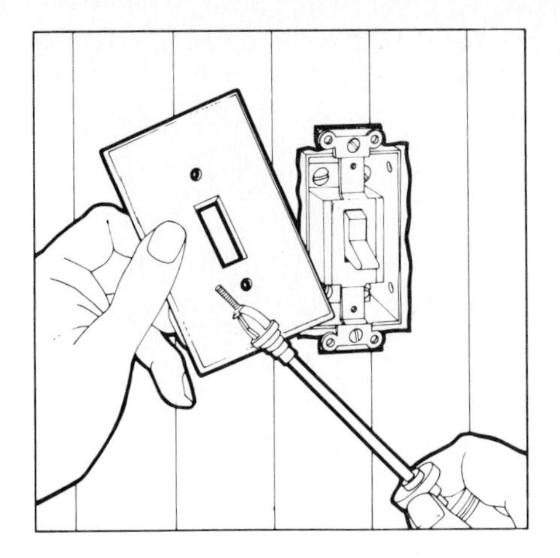

FIG. 8-20 Completion of the job after the inspection. (*Courtesy of AFC/A Nortek Co.*)

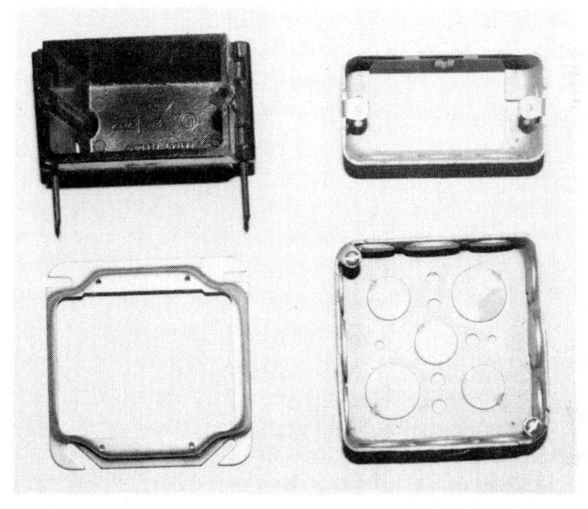

FIG. 8-21 Boxes used in wiring. (*Top*) Plastic wall box with nails for new work, box extension for handy box. (*Bottom*) Four-inch box cover which will provide space for two devices, 4-inch square junction box (J-box).

exception of the plastic wall box (upper left). Figure 8-22*a* shows a pull-in type of box anchor used in drywall construction when a box needs to be installed in finished walls. This is old work. Figure 8-22*b* shows anchoring items used for supporting boxes, conduit, and the like.

Cutting BX with a hacksaw is shown in Fig. 8-23. A closeup of BX including the plastic bushing is shown in Fig. 8-24. By using a

(*a*)

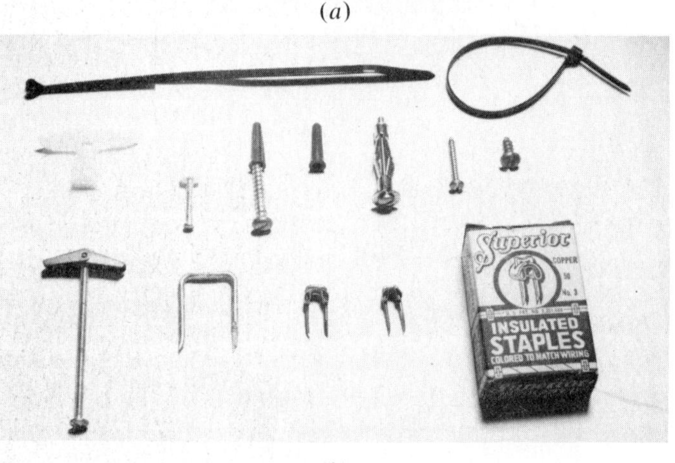

(*b*)

FIG. 8-22 (*a*) Back side of pull-in box support. Screw is tightened to secure box. (*b*) Anchoring materials. Top row: Nylon cable wrapping tie (T&B Ty-Rap), straight and looped. Center row: Plastic insert anchor; plastic wall plug with its own screw; wall plug and No. 10 sheet-metal screw, wall plug, Molly anchor, Nos. 8 and 10 sheet-metal screws. Bottom row: $\frac{3}{16}$-inch toggle bolt; service entrance cable staple; Nos. 7 and 3 insulated staples; box of staples (used for low-voltage wiring only).

box connector, BX can be brought into the box through a $\frac{1}{2}$-inch knockout (see Fig. 8-25). Make sure the box can then be inserted into the wall opening after the box connector is fastened in the box.

On occasion you may need a knockout opening in a special place in a box or panel. A punch-and-die combination called a *knockout punch* or a *hole saw* will make such a hole. Try to avoid this by using an offset nipple. Panels and large junction boxes have *concentric* knockouts—a series of knockouts, one on top of the other—to save space (see Fig. 8-26). These rings are concentric, one punched inward and the next punched outward. It is very

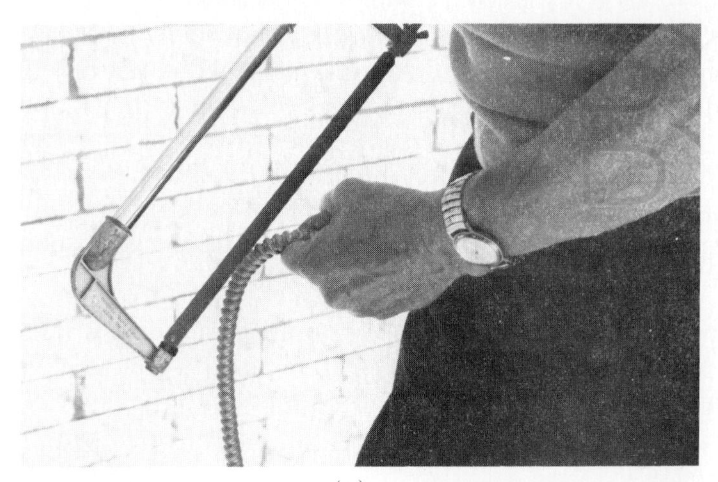

(*a*)

(*b*)

FIG. 8-23 (*a*) BX is sawed to remove covering. (*b*) The armor is removed, and the grounding strip is bent back. Next, the kraft paper is removed from around the insulated wires.

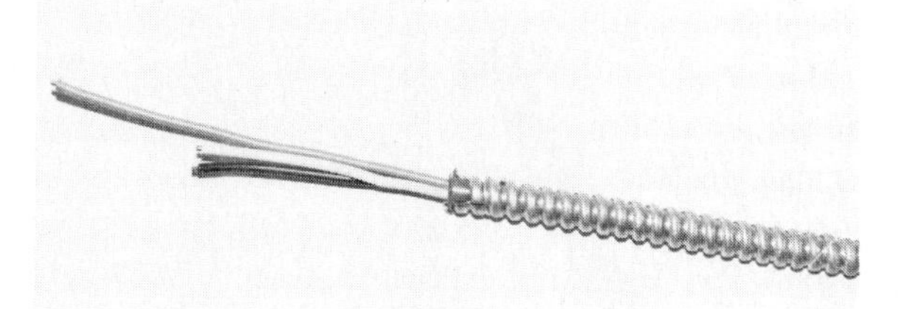

FIG. 8-24 BX is prepared for insertion in the box connector. Notice the anti-short-circuit bushing at the very end of the armor. This particular BX has an insulated green grounding wire, in addition to the metal grounding strip.

tricky to remove just the ones to end up with the correct size opening. As insurance there are always knockout reducers for sale. Be sure to punch in the right direction. The center may have to be cut with diagonal cutters, and the ends that are still connected are bent back and forth until they break off. Filing may also be necessary. The ring to be removed cannot be punched out all in one piece, but must be "worried" out with patience. Inspect such knockouts in a panel at a home-improvement center.

> **CAUTION:** Be sure the power is off if you are working in an operating panel. If one ring is left in place, a grounding locknut must be used with a jumper wire from it to the neutral bar.

WORKING WITH DRYWALL AND LATH AND PLASTER

Openings in drywall can be cut by only a hacksaw blade (wrap one end for a handle). Start with the blade nearly parallel to the wall surface. Saw on the box outline you have marked. The blade will

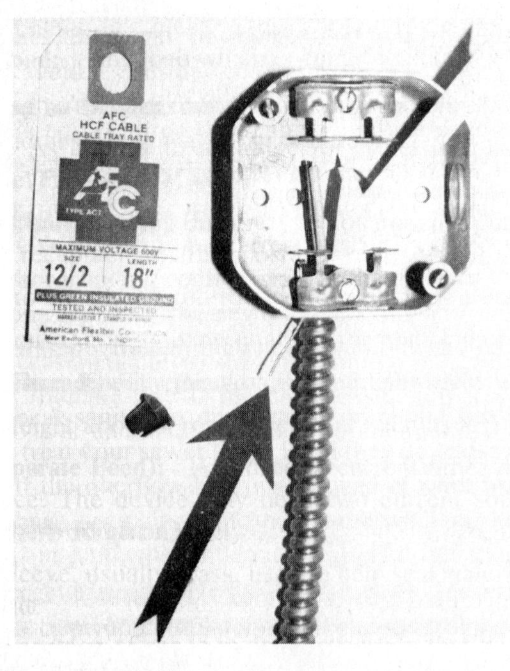

FIG. 8-25 BX inserted into the junction box. Anti-short-circuit bushing is shown in the space between the BX clamp and the bushing retainer.

(*a*)

(*b*)

FIG. 8-26 (*a*) Concentric knockouts in an entrance panel which allows for more knockouts given a smaller space. (*b*) The $\frac{3}{4}$-inch knockout has been knocked out, and the ring to enlarge the opening has been pried up.

sink right through the drywall. Saw each side in turn, leaving a little section not sawed; finish all sides. This method prevents breakage of the drywall in the wrong place. For lath and plaster, mark the box location but only on the sides (see Fig. 8-27). Chip away some plaster to expose one lath. This lath should be exactly in the center of the box side when the box is in the wall. If this lath would bring the box too high or too low, select a different lath. With the center of the box side on the center of this lath, mark the top and bottom of the box shape. Chip all the plaster from this area, using a small screwdriver with side cutters as a hammer. (This is against all rules for tool use but is done because plaster is soft and the tools will not be harmed.) This will expose an area the size of the box; one-half of the top lath, the full center lath, and one-half of the bottom lath will now be exposed. Very carefully and gently start to saw with your hacksaw blade only on the *top* half of the *bottom* lath. (You may have to angle the saw blade to insert it between the laths.) Saw both ends; with a screwdriver blade, try to split this piece from the lath. If the grain of the lath runs on a slant, try to pry in the other end. Twist the blade; *do not* pry sideways. When this piece has been removed, start on the whole center lath. Saw partway through each end. Hook your finger behind this lath to prevent it from moving while you are sawing (do not saw your finger). Plaster oozes through between the laths and runs down a little; this forms a "key" to hold the plaster to the laths. If the lath you are sawing wiggles, it may break off this key and the plaster may loosen and fall off in the wrong places. Finally, saw the *bottom* half of the *top* lath, and split that piece off. This should provide the approximate size opening. Trimming may be necessary, top and bottom, at the split surface. Space for the device mounting screws to enter the tab's top and bottom may need a little notching, as may the side where the box side is held by its screw. This is found on gangable boxes.

NEW WORK

When an attic or basement is finished, all the desired wiring can be installed before the finish wall is installed (drywall or paneling). This is called new work even though the building may be old. If you intend to build a recreation room in your basement, ask the electrical inspector whether there is a minimum height requirement for wall receptacles. Some areas subject to flooding may require that basement receptacles be at least 30 inches above the floor. This is to prevent live (hot) receptacles being submerged in water. This will not help your furnace or refrigerator, but may be required by local code. Many cities have local codes more restrictive than the NEC. Chicago will not allow nonmetallic sheathed cable. BX is required. In Detroit no BX is available in hardware stores.

Plan your wiring before you begin the work. Visualize the location of furniture and equipment, such as the stereo or TV. Plan to

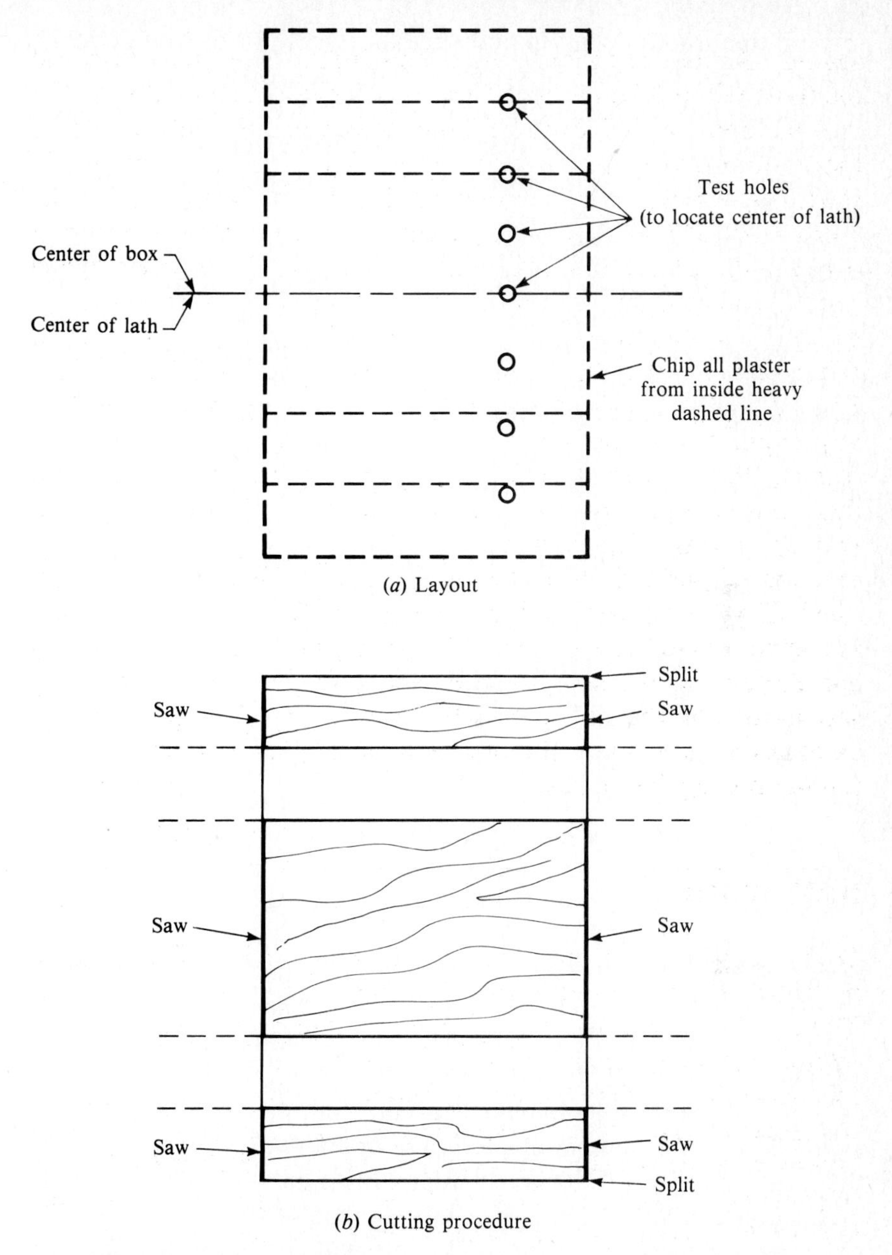

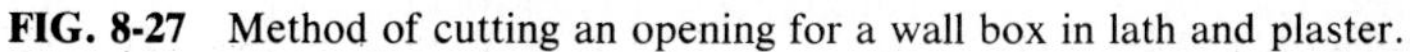

FIG. 8-27 Method of cutting an opening for a wall box in lath and plaster.

conceal the antenna lead-in and speaker wires just as you conceal power wiring. Homeowners can now do their own phone wiring. You may want to install phone jacks. The NEC, Sec. 800-3, requires that the separation of low-voltage from line-voltage wiring be at least 2 inches. Plan to keep a greater separation; in homes there is no need for closeness of these two systems.

The NEC requires that no wall space be farther from a receptacle than 6 ft (6-ft cords are standard). Also each wall space 2 ft or wider shall have a receptacle, and each counter space 1 ft or wider shall have a receptacle. Buy plastic or metal wall and ceiling boxes. Use those having nailing brackets for use on framing members. Position boxes *flush* with the finished wall surface, an NEC requirement. Plan to have two circuits if the entrancc panel has the space. A workshop should have a 20-A circuit wired with No. 12 copper wire. Lighting and small-appliance circuits can be fused at 15 A and wired with No. 14 copper wire. Present ceiling light wiring can be used if the box locations are suitable. The boxes can be reused, but the cable lengths may be short. Remember, all splices *must* be made inside a junction box, and the box must have a cover and be accessible without damage to the building construction.

Be sure not to exceed the bulb size in new fixtures; this is also an NEC requirement. Flush fixtures are a special case because they can become very hot. Special wiring is used and brought out of the fixture into a junction box which is a part of the fixture assembly. Your wiring must be connected to the wires in this box. Notice that all these NEC requirements are concerned with safety of persons and fire prevention. *This* is what the NEC is all about.

Visit your hardware store to pick out the necessary boxes and other items. Ceiling boxes should have the adjustable bracket to use between joists. This is the most satisfactory type. Work neatly as you run the cable lengths. If you plan to install three-way switches at the basement or second-floor stairways, remember that the cable *between* the two switches must be 14-3 with ground, either nonmetallic or BX cable. Use a piece of cord or clothesline to thread through where the 14-3 cable will go; run it as if it were the cable. Remove it and measure its length. This will tell you how much cable you will need to buy. Add 10 percent for errors in measuring; if the line measures 15 ft, buy 17 ft. The other cable should be bought in lengths of 50–100 ft. Three-way wiring and cable are shown in Fig. 8-28.

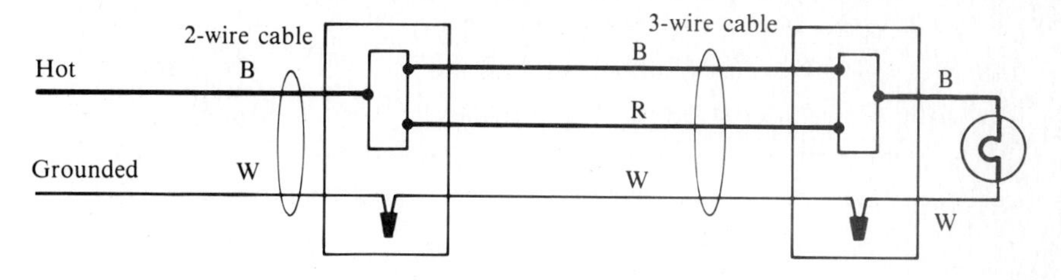

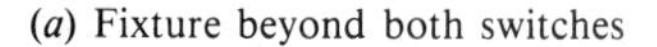
(*a*) Fixture beyond both switches

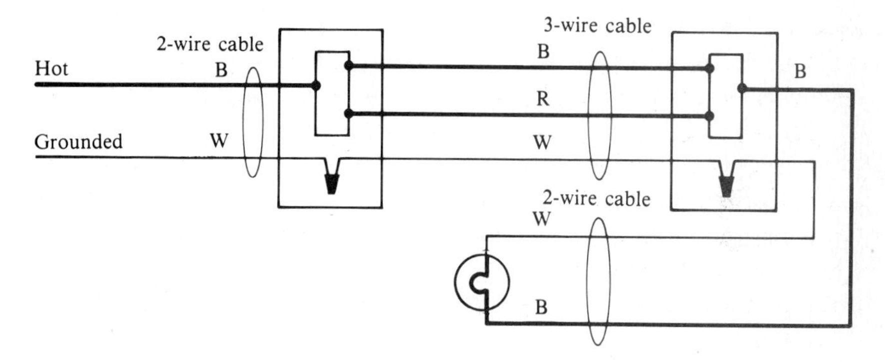

(*b*) Fixture between both switches

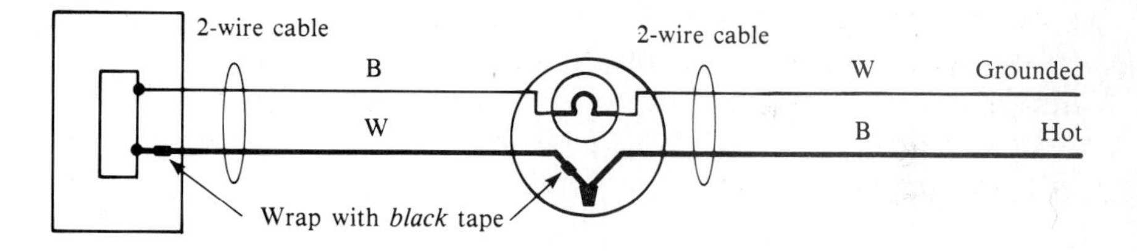

(*c*) Use of white wire as a hot wire (NEC Sect. 200-7, Exception No. 2)

FIG. 8-28 Wiring of three-way switches using 14-3 with ground (ground wire not shown) between the switch boxes. Also shown is use of 14-2 with ground Romex as a switch leg. Each end of the white wire is wrapped with tape because it is hot. (See NEC, Sec. 200-7, Exception No. 2.)

Route and secure the cable as neatly as possible. This will impress the electrical inspector also. Leave at least 8 inches at junction boxes and wall or ceiling boxes. If the cable is to originate at the main panel, make sure the leads there are long enough; you may need 2 ft or more inside the panel to make connections. Use the staples having the plastic bar across the top with a nail in each

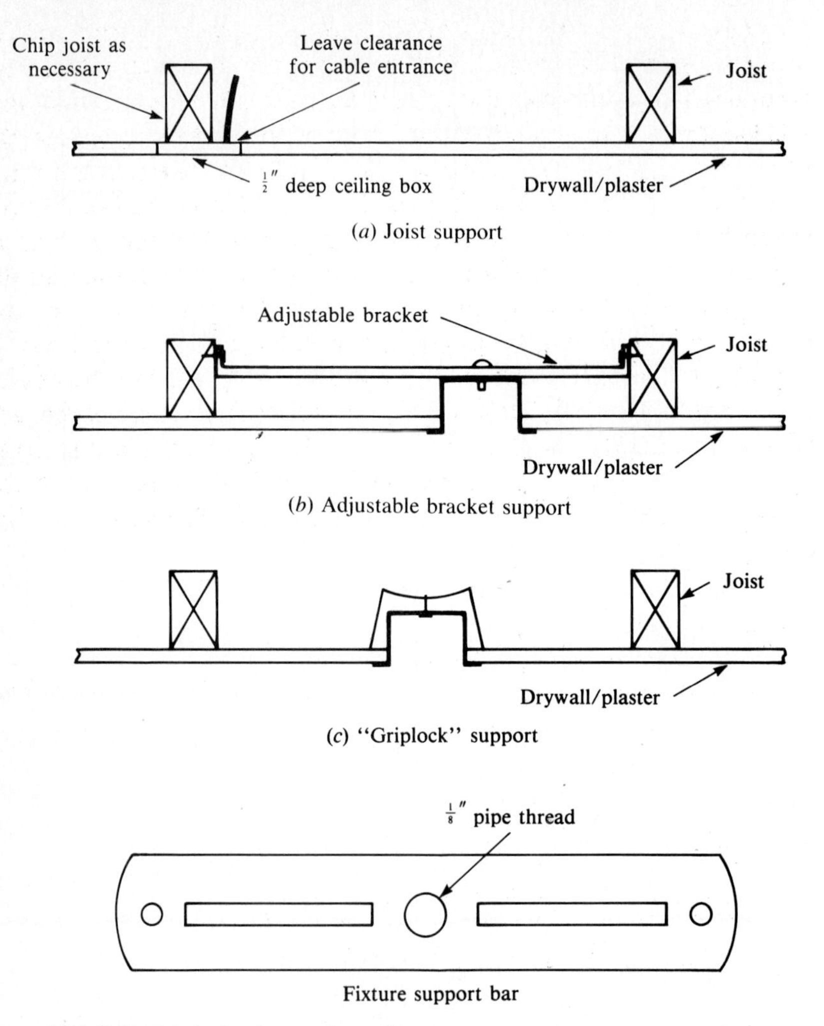

FIG. 8-29 Methods of mounting ceiling boxes with various types of support.

end. Complete all the additional work before connecting anything to the main panel. Then you can work with no rush and no disruption of power to the system. The final tie-in can be done in about a half-hour. Power will be off only during that time.

> **CAUTION:** Before you make the final connection, check all your work for proper connections and tightness of terminal screws. Be sure no bare wires are exposed.

Most electricians tape the skirt of wire connectors (wire nuts) for protection of bare wires and to prevent the connector from becoming loose. Before taping, check that the connector is tight and pull on each wire in turn to see if any wire is not caught in the wire connector. Neatly fold the wires back into the junction box. Curl each pigtail in a half circle so it will lie flat in the box. Avoid sharp bends. Tape the terminals on switches and receptacles to prevent short circuits. Start the tape at one screw; continue to the other screw and around the device body 3 times. Push the bare ground wires and their wire connector as far back into the box as possible. Make an S curve of the hot wires as you push the device into the box. With the service entrance panel energized but the new wiring *not* connected, you can work safely and carefully to complete a proper installation. Then make the final connection to the entrance panel. Ceiling box mounting methods are shown in Fig. 8-29.

Small-Appliance Repair

PLUG AND CORD REPLACEMENT

Repairs to small appliances usually consist of the replacement of either the attachment plug on the cord or the complete cord and plug. Some newer appliances including electric drills have detachable cords which can be easily replaced.

The common molded-on attachment plug is subject to much strain when it is inserted or removed from the receptacle. Most plugs are hard to grip without grasping the cord also; thus most of the pull is on the cord rather than on the plug. The wires from the cord are usually only crimped to the inside end of the plug prongs; any strain on the wires will loosen them from their connection. This will cause overheating of the plug prongs and body and cause more overheating. Hard plastic plugs will crack and may flash sparks. In many cases cutting off about 6 inches of cord and installing a new plug cures the trouble. The new plugs now have dead-front construction (no exposed live parts to contact metal receptacle plates).

Inspect the total cord length for worn or frayed spots. The appliance end and plug end of the cord are subject to cracking from the heat of the appliance or the overheated plug (rubber cords become hard and brittle when they get too hot). *Any* evidence, no matter how slight, of damage to the cord calls for replacement of both cord and plug. Replacement cords are sold in hardware stores.

> ***Note:*** Heating appliances such as toasters, irons, and waffle irons must have a type of cord designated *heater cord.*

This is heavier than lamp cord and should be No. 16 AWG at least. Certain replacement cords will have ring ends (loops) on the appliance end; others will have stripped ends.

Appliances are assembled in various tricky ways, and you will have to determine how to open the case to get at the cord end. Some have screws or rivets; others snap together and must be pried apart (*after* you have found where to pry). Be very careful when disassembling them not to damage any parts. The cord will enter the case through a bushing or other smooth, rounded surface. Newer appliances feed the cord through a clamp arrangement which deforms the cord into an S curve; this effectively keeps it

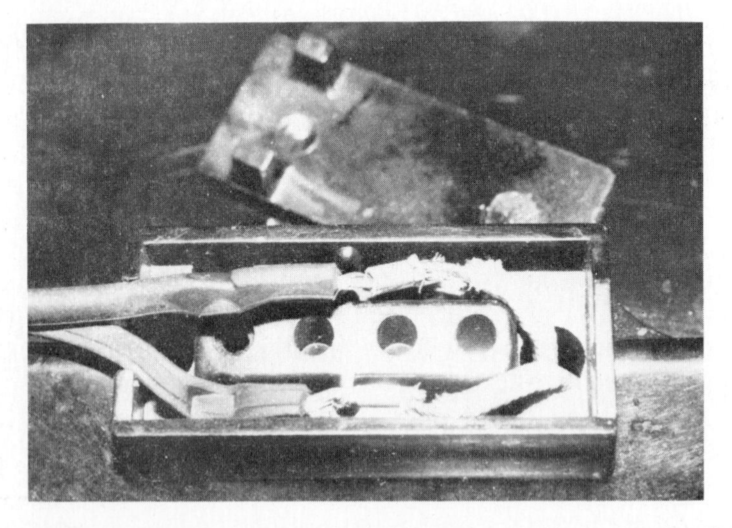

FIG. 9-1 Strain-relief clamp. Note how wire is deformed to provide the clamping effect. In the center are shown crimp-type wire connectors joining the cord to the interior wires. Cover is above connections.

FIG. 9-2 Older style of strain-relief clamp using disk and cord wrapping.

from moving and thus keeps the strain off the terminals. Some of these crimping clamps are very hard to remove; they have to be squeezed with Channellocks very tightly while they are removed from the opening in the appliance case. Figure 9-1 shows a waffle iron having a modified pressure grip similar to the one just described. The cover plate has two bosses to press each wire down in the groove. Also notice that the electric connectors are made with crimp-type connectors. Figure 9-2 shows the cord strain-relief device of a Toastmaster Model 1B14 toaster. This device is a fiber disk on the cord; beyond that is a winding of thread or light cord tied to the appliance cord. This resists the strain on the cord from being pulled out of the case. This old asbestos-type cord can't be deformed as the new rubber cord can, so this method is used.

Every manufacturer designs products differently from other manufacturers. Newer appliances are more complicated and difficult to repair. Refer to Figs. 9-3, 9-4, and 9-5. The waffle iron has replaceable heating elements which are supported on porcelain; these are coiled nichrome alloy wire which can be stretched to the length needed. Carefully remove the broken element so as to be able to measure its length. Most hardware stores have these elements. Get the same wattage. The element shown is 1200 W. This iron has a thermostat. The thermostat will have to be purchased from the manufacturer's service repair center, in this case, General

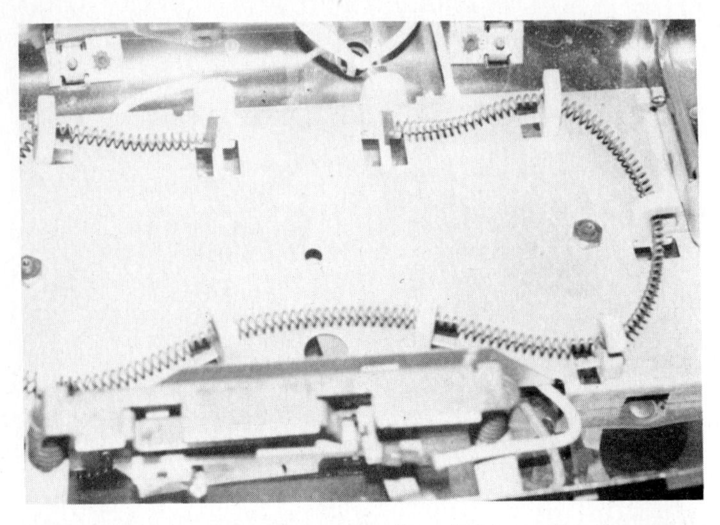

FIG. 9-3 Interior of a waffle iron showing coiled type of heating element.

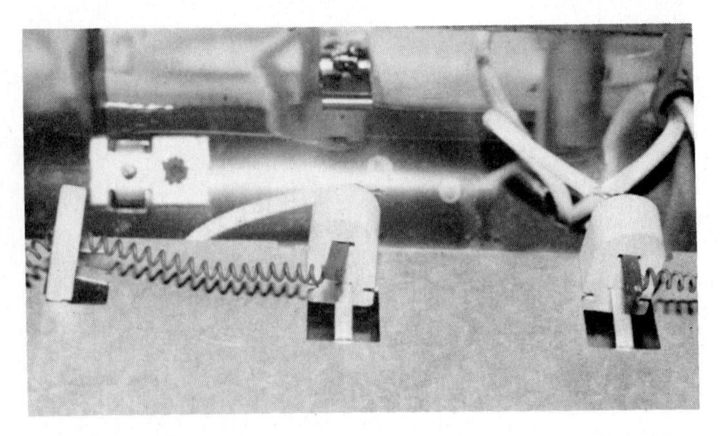

FIG. 9-4 Closeup of the heating-coil connections to the interior wiring.

Electric. Large cities have independent parts distributors who may stock identical or replacement parts such as this thermostat.

These parts suppliers also carry replacement parts for electric ranges: surface elements, oven elements, burner switches, and thermostats for the oven. The counterperson is knowledgeable and will help you select what you need. Newer appliances may have electronic circuits which will need special equipment to test them for defects. It is best to let the service center repair these appliances.

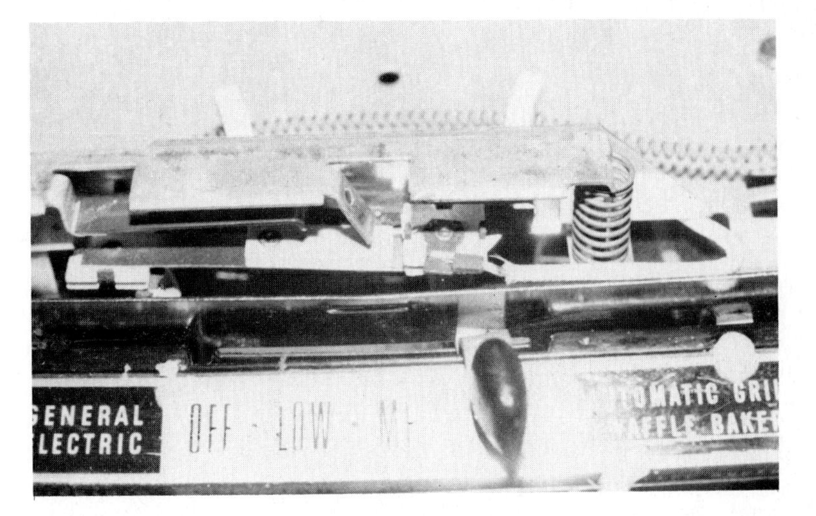

FIG. 9-5 View of a waffle iron thermostat showing adjusting lever at the front of the appliance.

Table and floor lamps are easy to fix. They consist of cord, plug, and socket. Lamp cords do not get the wear that appliance cords get. The lamp socket gets the wear; it may be three-light, on-off turn, or on-off push-through. Most are three-light. To replace the socket, unplug the lamp. Remove the shade and bulb. Near the switch knob will be the word *push* on the brass-plated socket cover (shell). Push here, and the two parts should separate. If they do not, insert a small pocket screwdriver between the two parts and pry gently sideways (see Fig. 9-6). Pull the shell off the socket mechanism. Feed some cord into the lamp base so as to allow the socket to be pulled up slightly, so you can get at the screw terminals. Inside both the shell and base are fiber liners as insulators (see Fig. 9-7). Do not damage or lose them. Purchase a replacement socket at the hardware store identical to the original. Loosen the terminal screws, and remove the old socket. Reverse the process to install the new socket. Appliances such as this one do not follow the rule of black wire to brass screw and white wire to white screw because the attachment plug may be inserted either way in the receptacle. Certain electronic equipment may have *polarized plugs* (one prong is wider, and so the plug can be put into the receptacle in only one direction) but lamps, toasters, etc. do not have them. Be sure to twist the stranded wire of the cord tightly before wrapping it around the terminal screw. Tinning the ends of stranded wire makes it easier to get the strands to stay in place. If you do this, use *rosin-core* solder and a soldering gun or iron. Wires are always wrapped clockwise around any terminal screw whether stranded or solid wire. This tends to keep the wire from unwrapping as the screw is tightened. If the wire is not tinned, be sure to check for stray strands sticking out from under the screw.

Vacuum cleaners have a special cord listed in the NEC as "vacuum cleaner cord." This is a round cord, No. 18 AWG. The cord attaches to the switch, usually in the top of the handle. Sometimes a cleaner comes with a cord just too short to clean a living room without changing receptacles. About 5 ft longer would be ideal. While the NEC does not prohibit extension cords, it discourages their use as a permanent item. To replace the cord, remove the switch screws and pull out the switch. The new cord does not *have* to be vacuum cleaner cord, but should be a good grade of round rubber-covered cord. Since this cord will have the black and white wires, connect the black wire to the switch terminal and the white wire to the white wire going to the motor. The white motor wire may be connected to the old cord by a crimp connector. Cut this off, and use a small wire connector. When you repair any appliance, it is very handy to have a terminal crimping tool. Many connectors used to tie two wires together are too small to be able to fit in places such as this switch area. Screw-on wire connectors are larger and may not fit here.

ELECTRIC MOTORS

Electric motors can be disassembled, cleaned and oiled, and reassembled easily. Types of motors you can service are found on furnaces, washers, dryers, portable fans, and pumps. All are basically alike. Inside the motor housing is a centrifugal switch operated by weights attached to the rotor, the rotating part of the motor. The switch contacts are closed when the motor is stopped and until it attains full speed. When the motor reaches full speed, the weights spread out and by means of levers push the switch *open;* this disconnects the *starting* winding of the motor. The other winding, the *running* winding, stays connected, and the motor stays running until it is shut off. Upon restart, the cycle is repeated. This motor is called a *split-phase* motor because the two windings react

FIG. 9-6 Typical lamp socket (three-way) showing method of disassembly by using a screwdriver.

FIG. 9-7 Component parts of a typical lamp socket.

to start the motor under a load, such as a washing machine going into the spin cycle.

To disassemble the motor completely, disconnect it from any power source. The motor housing consists of two end bells and the center section. The shaft (front) end bell has no internal parts except the bearing. The rear end bell has a terminal board and the centrifugal switch. Using a prick punch, make a punch mark on the edge of the rear end bell. Make an identical mark on the center section directly opposite the end bell punch mark (see Fig. 9-8*a*).

(*a*)

(*b*)

FIG. 9-8 (*a*) Method of marking an electric motor before disassembly. Arrows point to prick punch marks. (*b*) Start of disassembly by using a screwdriver to start removal of end bell.

This will give you a reference point from which to reassemble the end bell. Do the same with the other end bell, except make *two* marks on the bell and the center section also. You now have *no* excuse for putting the motor together wrong. You can now remove the through bolts; these will have a screw head on one end and a nut on the other (see Fig. 9-9). Remove all four. The joint where the end bell meets the center section is a press fit. There may be small slots around this joint, so one can insert a screwdriver and tap with electrician's sidecutters (see Fig. 9-8*b*). If not, there may be humps on the edge of the end bell to tap against. Tap at different

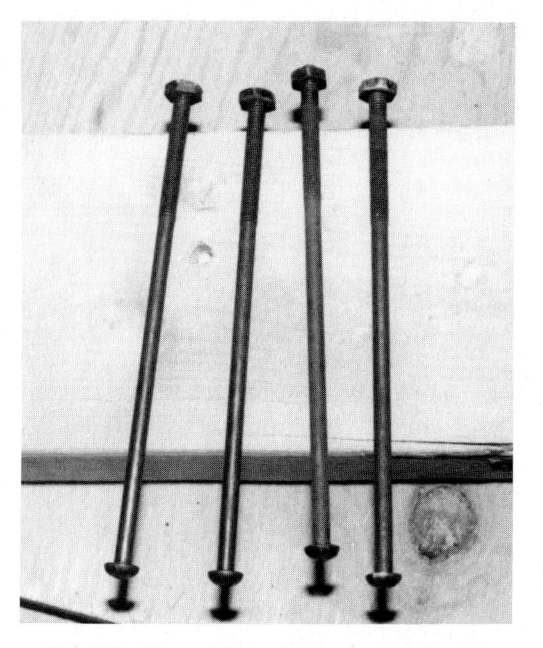

FIG. 9-9 Through-bolts from motor in Fig. 9-8.

points around the circumference, because the bell must come free straight out and not at an angle. Some are easy; others are tight. Tap gently, for these parts are thin castings. Lay the bell close to the center because there will be wires going to the starting winding. You will now be able to see the switch contacts (see Fig. 9-10). They will be dirty from dust and oil. Shine them, but do *not* file them. Small acid brushes, ½ inch wide with tin handles, make fine dust brushes for these jobs. Some hardware stores have them.

If enough dirt gets on these contact points, they will not pass current and the motor will not start. It will hum and trip its overload switch; if it is automatic, the overload will reset and try many times. If the motor is manual, it will need to be reset (pushed in). If the motor then hums, you know the switch is dirty. The automatic reset overload will not be noticed until conditions resulting from the motor not running are noticed (no heat from the furnace, for example). In my case, a sump pump motor was trying to start and dimming the lights. A business card drawn through the contacts cleaned them easily (see Fig. 9-11). If the points are burned badly, go to the parts distributor for an identical replacement. (Take the old part with you to ensure the proper match.)

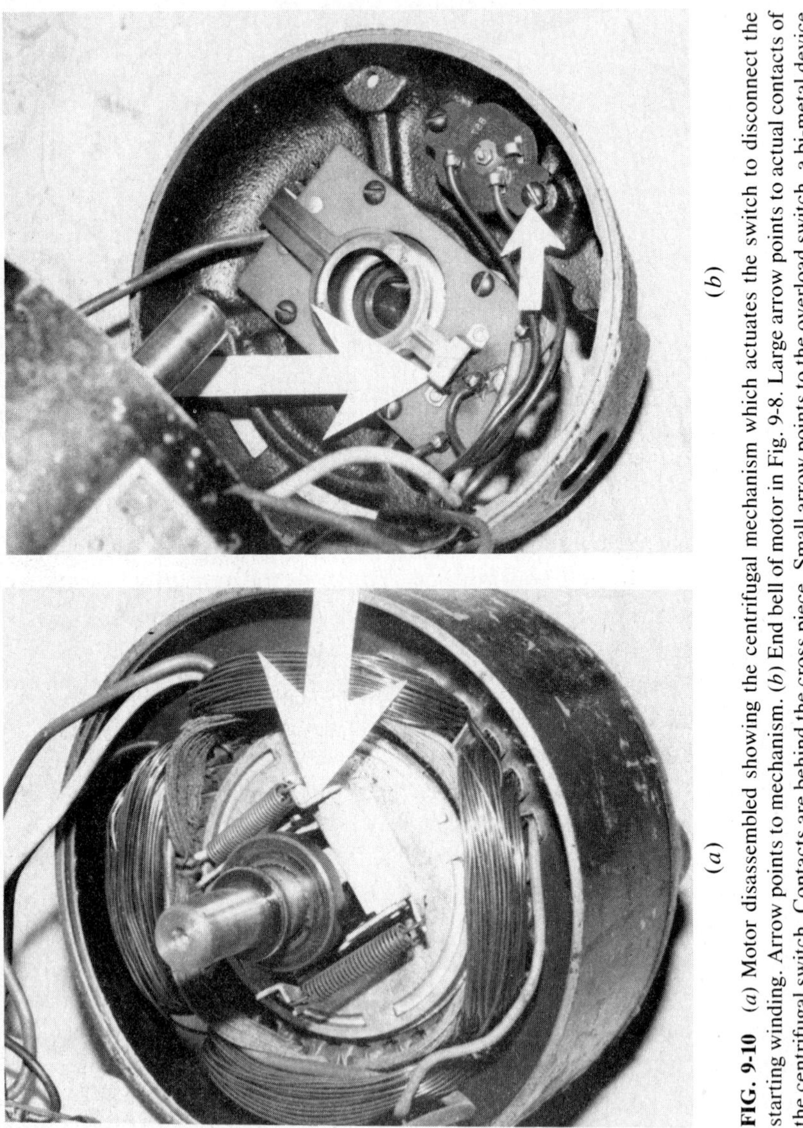

FIG. 9-10 (*a*) Motor disassembled showing the centrifugal mechanism which actuates the switch to disconnect the starting winding. Arrow points to mechanism. (*b*) End bell of motor in Fig. 9-8. Large arrow points to actual contacts of the centrifugal switch. Contacts are behind the cross piece. Small arrow points to the overload switch, a bi-metal device actuated by the heat of the motor.

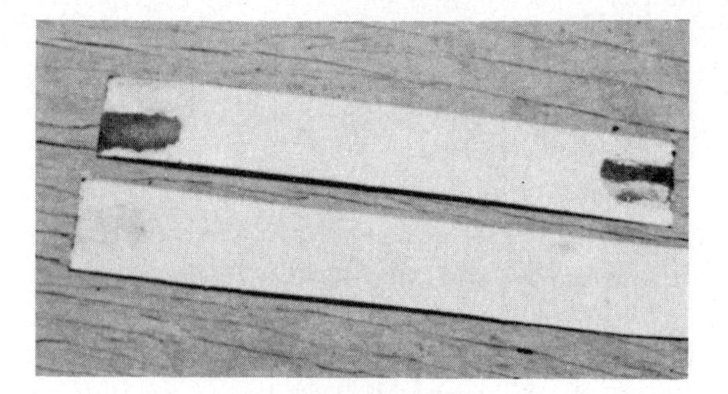

FIG. 9-11 (*Top*) A strip of business card is pulled through the contact points of the motor in Fig. 9-8. (*Bottom*) Lower strip shows the contacts are now clean.

FIG. 9-12 Exterior of motor end bell. Connection terminals are shown at the top and exterior of the motor overload switch.

Because of the dust accumulation on the contacts of the centrifugal switch, this type of motor must be oiled very sparingly—four or five drops twice a year, at the start of the heating season and in January. Motors such as these have a large wool wicking in a recess around each bearing. This will hold a quantity of oil in reserve. The above oiling frequency is for furnace blower motors but will hold true for most motors which do not run continuously. Figure 9-12 shows the terminal connections and the automatic overload location.

I once found a furnace blower motor from which I drained at least 2 Tbsp of oil. This oil was in the winding area and not in the bearings.

CAUTION: Do not use detergent-type car motor oil. Use either a No. 10 or No. 20 weight motor oil *nondetergent-type* or an oil recommended for electric motors.

Any motor when *starting* will draw from 2½ to 3 times its normal *running* current. This is characteristic of all motors. This current draw rapidly diminishes as the motor comes up to its normal speed. If a motor blows fuses often, the fuse should be replaced with a time-delay fuse, previously described. This fuse will hold for the high-current inrush during the short time needed but will blow when necessary because of a short circuit or other fault. These time-delay fuses are available everywhere. Be sure to replace your fuse with the same size time-delay fuse. Do *not* use a larger-size fuse.

CHAPTER 10

Special Projects

Many times a small outlay of money and time results in improved comfort and convenience for the occupants of a dwelling. In other cases there is a saving in energy costs for heating and cooling.

NIGHT SETBACK THERMOSTAT

The night setback thermostat is a simple project for the homeowner. The cost of a top quality thermostat is about $75. The time to recover this cost is, in some cases, as little as 2 years. Most companies making thermostats also have a *clock thermostat*. The clock part operates the switching device, which lowers the temperature during the night and raises it again in the morning (see Fig. 10-1). You, the homeowner, select the times for these actions and the amount of temperature lowering. People with small babies may not be able to select as great temperature lowering as those in a home of all adults. It is estimated that one 10-degree setback will give a savings of 11 percent while two setbacks (one during the day when no one is home) will give a savings of 13 percent. This is in cities such as Chicago and Detroit.

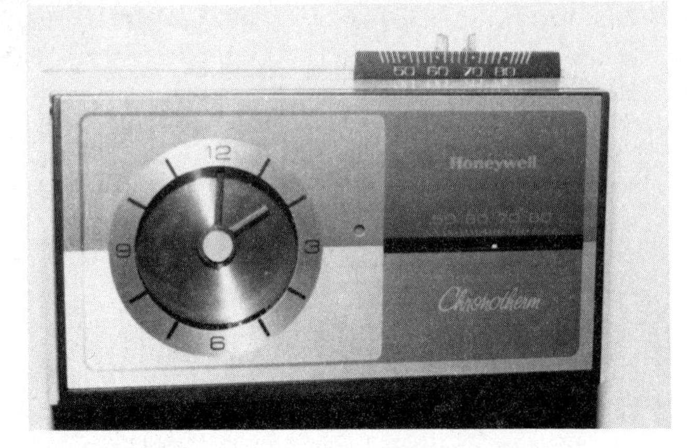

FIG. 10-1 Night setback thermostat. The two temperature setting pointers are on the top.

As with all purchases, only buy items on sale. A recent advertisement showed a heating-only thermostat at $50 and a heating/cooling thermostat for $60. All major brands are reliable and cost about the same unless the store has a sale going on. List price for Honeywell thermostats can be as high as $200 for night setback, but are usually available for under $100. Because of the clock, all thermostats must have continuous power to actuate the temperature change. Wiring diagrams are included with each thermostat. All power to operate thermostats is only 24 V. Be sure to trip the breaker or remove the fuse so that all equipment is off. For heating or cooling a subbase is needed. This accounts for the difference in the prices quoted before. This subbase provides the selection levers needed to switch from heating to cooling, or the reverse, and to provide for continuous or automatic fan operation.

After shutting off the power, remove the old thermostat. The cover is held on by springs or a wedge fit over plastic knobs (see Fig. 10-2*a*). Most thermostats have a mercury bulb for the contacts; yours may not. Nonetheless, there will be two or more round $\frac{1}{8}$-inch openings with a screw slot showing in behind. These screws are *captive* and hold the thermostat to its mounting plate. Using a small screwdriver, loosen all these screws. They will not come completely free, because they are captive. Hold the thermostat before loosening *all* the screws. This will leave only the mounting plate. The thermostat wires from the equipment are connected to this plate. As you disconnect these wires, be sure to identify each one with masking tape or otherwise mark them with the designation shown on the base near each terminal. Now, remove the screws holding the base to the wall, and carefully pull the base free, guiding the wires through the base opening. The new thermostat either will come with the extra subbase or, if for heating only, will have just its mounting plate. The installation is the reverse of the old thermostat removal.

(*a*)

(*b*)

FIG. 10-2 (*a*) Inside view of a thermostat showing the 24-hour dial with the changeover fingers in place at: off—11:30 PM, on—7:00 AM. (*b*) Interior view showing the mercury tube and adjusting potentiometer. Setting must match amperage of gas valve or valve relay.

Note: Be sure to plug the wall opening where the wires come through because air currents can enter here and affect the thermostat's correct operation.

Follow the directions for moutning the subbase or mounting plate, using suitable wall anchors and screws. If the wires are long enough, strip enough insulation to form new ends (sometimes the old loops will break off and cause a no-heat condition that may be hard to find). Attach the wires to the proper terminals according to the diagram appropriate to your system. Mount the thermostat by setting it on the base and tightening the captive screws. All thermostats have inside, near the mercury bulb, a scale with a movable pointer (see Fig. 10-2*b*). If the old thermostat had a similar scale, set the pointer to the same figure, which will be 0.2, 0.45, 0.8, or a similar figure. If the old thermostat has none, you will need to find the amperage rating on the gas valve, oil burner, primary relay, or equivalent device. Set the pointer at this figure. Now install the cover. Restore the power, and try out the system. Most instruction books have detailed instructions and helpful hints for troubleshooting. Just make sure you wired the thermostat correctly. Finally, follow the instructions for setting the 24-hour clock tripping pins for the desired setback and setup times.

DOUBLE-THERMOSTAT ARRANGEMENT

This night setback arrangement uses two thermostats—the original thermostat and one heating-only thermostat. In addition, a timer for switching is needed. This timer disconnects the day thermostat (original thermostat) from the furnace during the night setback period. When this happens, the night thermostat takes over; as this thermostat is set perhaps 10 degrees lower, the house cools down to that setting before the furnace starts. In the morning the reverse happens, and the day thermostat is already calling for heat. A heating thermostat costs about $30, and a timer, $40; this is cheaper than a clock thermostat.

The extra thermostat can be located wherever you desire as long as it is in the general area of the original thermostat, to sense the same temperatures. Put the timer in the basement near the furnace. The timer needs line voltage to run its clock.

Note: The NEC prohibits low-voltage wires in the same enclosure as line voltage, 120 V.

To comply with this rule, mount a 4-inch box next to the timer case with an offset nipple. Run No. 14 black wires from the timer low-voltage contacts out and into the 4-inch box, leaving enough for connections, and run the low-voltage wires into the box and make the connections. The No. 14 wires can be taken from a piece of Romex—2 ft is plenty. These wires are line-voltage-insulated and are OK to use. Figure 10-3 shows how to wire the two thermostats and the timer into the heating-only circuit.

Note: Most list prices are discounted greatly. Shop around for prices.

AUTOMATIC GARAGE DOOR OPERATOR

Garage door operators have become almost a necessity because they are such a convenience in bad weather. All companies market easy do-it-yourself kits with detailed instructions. Most garages already have a receptacle close, especially for the operator. If there is none, you will be able to use your knowledge and skill to install a new circuit (it must be on its own separate circuit) from the service panel to the garage location. Actually run the circuit from the garage, and make the panel connection last. Mount the push-button near the house door for convenience.

Adding a Door-Open Indicator

Many times the garage door is left open by mistake; this makes it easy for thieves to steal from the garage. Adding a pilot light inside the house alerts persons to the open-door condition. A second pilot light can be added to indicate that the garage ceiling light is on. Both lights can be made one project. Required are a wall box, duplex receptacle, two plug-in night lights, and sufficient Romex to wire these lights. Most door operators have a light which comes on when the door is open and stays on until the door is closed, You can tap into the two wires (one black, one white), which connect to

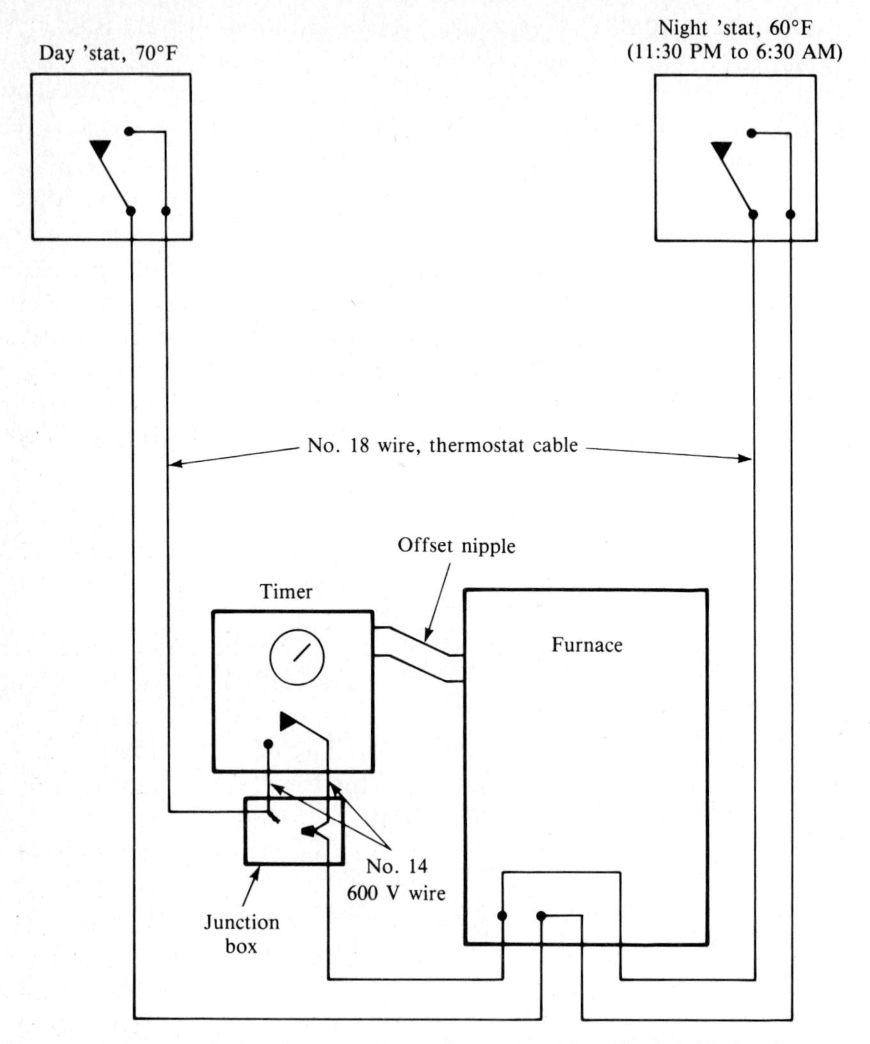

FIG. 10-3 Wiring diagram of a two-thermostat night setback (with timer).

the light socket or eliminate the light and tie directly to the contacts where the light wires were connected. Some lights come on when the door opens but go out a short time later (5–10 minutes). In this case you will have to install an end switch on the side of the door track to actuate the pilot light circuit (or bypass the switch which turns the light out). The ceiling light pilot is connected to the ceiling light wires in the junction box at the light itself. These connections and diagrams are shown in Figs. 10-4 and 10-5.

Mount the duplex receptacle inside the house where it is easily seen. The receptacle must be modified by breaking out the connecting bars on both sides of the receptacle—both the brass-screws side and the white-screws side. This isolates the two outlets completely, which is what you have to provide. Romex is used to run both circuits, one from the door operator and the other from the light fixture.

Either a ceiling or high wall location is ideal for these lights, so they can be seen readily. I have mine directly above the door leading to the garage. The night lights are clear plastic, roughly triangular, and have a red neon light inside. They have prongs which plug directly into the receptacle. When either light is on, both conditions can be checked at one time, since both are in the garage area.

OUTDOOR PERIMETER LIGHTING

Perimeter lighting is becoming more necessary because of the increase in vandalism; it is not too difficult to install. By using thinwall tubing and installing floodlights on each corner of the building, you will have the perimeter well lighted. If desired, decorative lighting may be used instead. This method does not give as bright light but will still be a deterrent.

The corner floodlights will require weatherproof boxes and covers which will accept adjustable sockets for the floods. Thinwall with connectors which screw into threaded openings in the boxes help make a waterproof installation. If your house has eaves all around, this work will be easier. At one of the corners you can bring a supply lead out from the service panel to furnish power to the lights. You may need a junction box to support a photocell to control the lighting. Photocells must be oriented to avoid lights which will affect their sensing of darkness and make operation erratic. If you can, use this same box to bring the feed from the panel, and make all connections in this one box. Use a $4\frac{11}{16}$-inch-square by $2\frac{1}{2}$-inch-deep box.

The individual lighting fixtures will be more difficult to wire because you will be bringing Romex up inside an outside wall filled with insulation (usually). Perseverance will pay off if you want this style of light, but you will have a more difficult job. You can drill

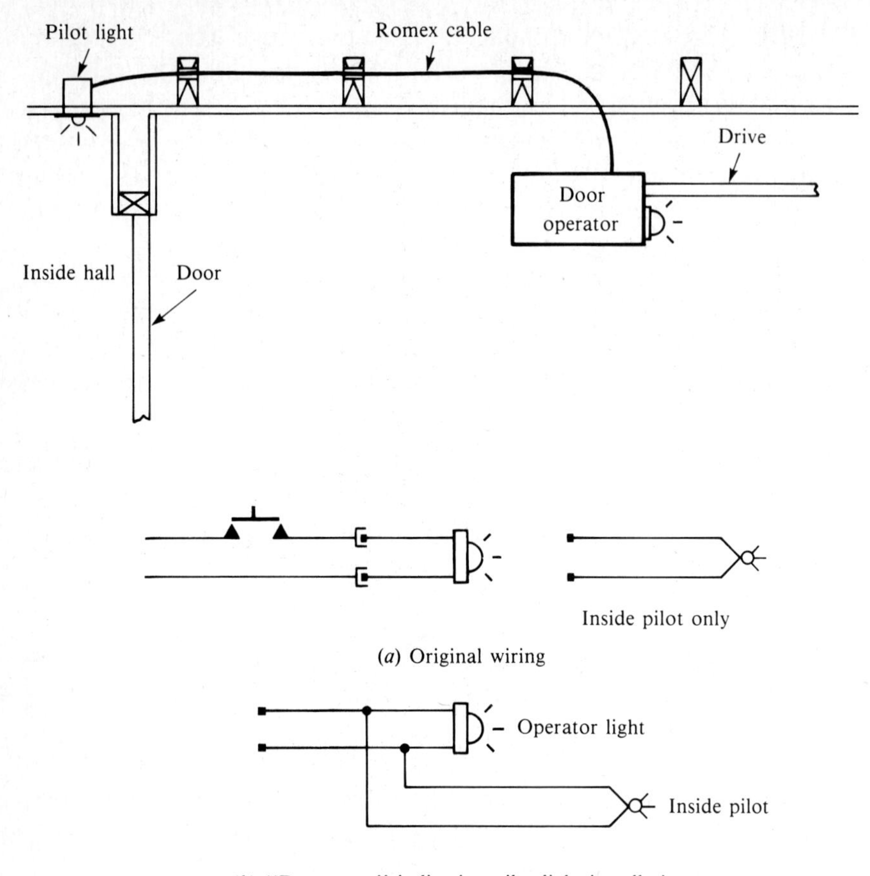

(*a*) Original wiring

(*b*) "Door open" indicating pilot light installation

FIG. 10-4 Layout and diagram of a door-open pilot light.

through brick veneer to bring the Romex through and use a ½-inch-deep round box to connect the fixture and hold the splices necessary. I did one house where the lights were on each side of the large garage door and one was on the side of the garage near the front house entrance. These walls had no insulation, and so it was easy to run Romex through them. The photocell was on the opposite side of the garage to avoid street lights and cars going by. Buy a good-quality photocell; the cost is $10. A timer can be used, but it costs more, about $40, and has to be reset to accommodate the changes in hours of darkness. The layout for outdoor lighting is shown in Fig. 10-6. Note that floodlights can shine in two directions from each corner of the house.

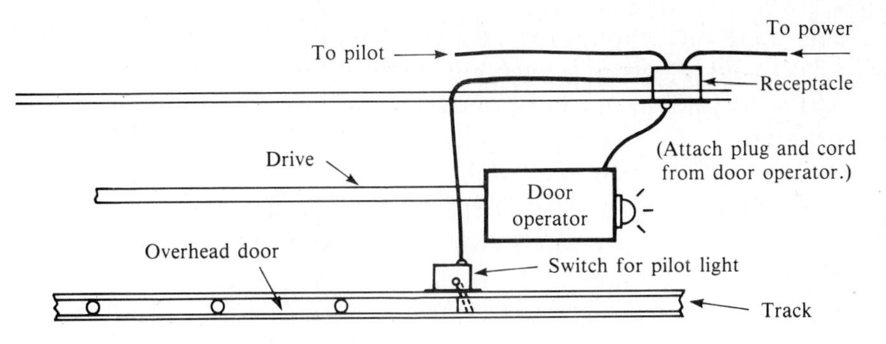

(*a*) Location of end switch for pilot light

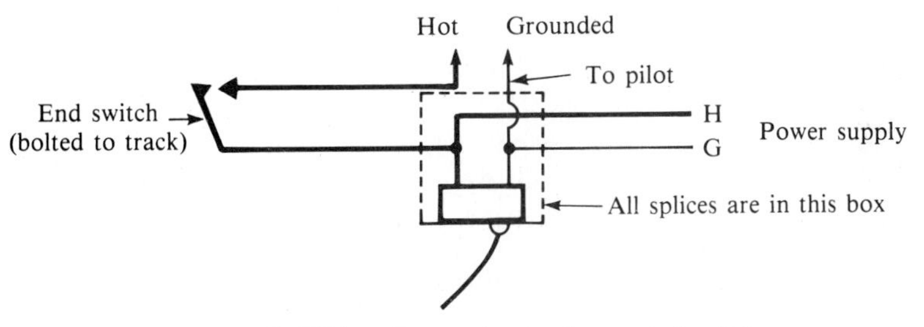

(*b*) Wiring diagram for track-mounted switch

FIG. 10-5 Alternative layout and wiring method when door operator light goes out after a short time.

ELECTRIC WATER HEATER CONTROL WITH A TIMER

Your electric water heater can be controlled by a timer to save electricity. The timer will cost $45–$50. The timer must have a capacity to handle 40 A, so be sure to check the current-handling capacity. The clock motor can be either 120 or 240 V. Refer to Figs. 10-7 and 10-8.

> ***Note:*** Certain utilities have an arrangement for shutting off the water heater to reduce the load on their lines at their discretion. Be sure to check with your utility company to find out if you have interruptible power before you buy the timer.

The utility determines the off times by calculating their connected load. A high load calls for the turning off of certain "interruptible loads" in an attempt to equalize the total connected load on their generators. If you install a timer on a utility-controlled water heater, you will need to supply the clock of the timer from a separate current source. The clock must run continuously to maintain correct on and off periods, which you have selected. I find that 2 hours in the morning and 2 hours in the evening are sufficient for my household. Other families may need longer on periods. Other utilities feed the water heater through the regular meter, and *no* power interruption takes place; in this case buy a timer with a 240-V clock. If you have the power interruption system, buy a timer with a 120-V clock; it is easier to supply 120-V power from the entrance panel since a line can be run from any continuously powered circuit. Since the clock takes less than ½ A, any circuit can be used.

The water heater will have greenfield from a wall box or directly out from the wall to the heater junction box. In basements Romex will come from the ceiling down the wall inside thinwall tubing ending in the box. The Romex cable covering has to be removed to fit in the thinwall. From the wall box or directly out from the wall, greenfield leads to the heater junction box. You can mount the new timer next to the junction box by using an offset nipple; use a ¾- or 1-inch size. If there is plenty of length on pigtail connections in the wall box, run them into the timer case (the

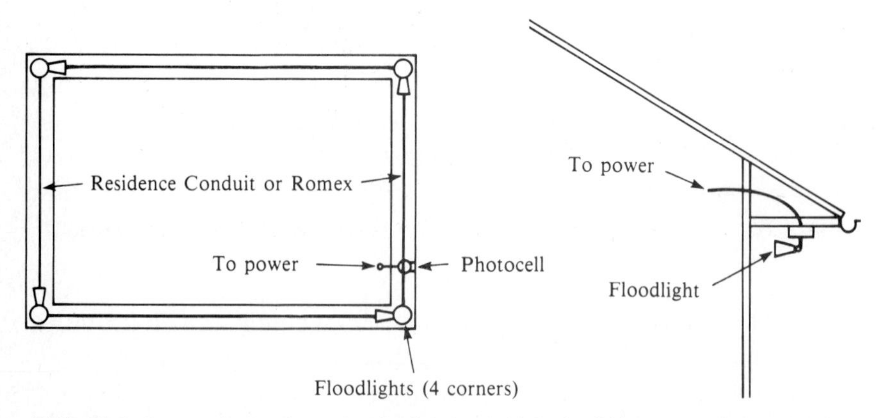

FIG. 10-6 Layout for outdoor security lighting (which should be on a separate circuit).

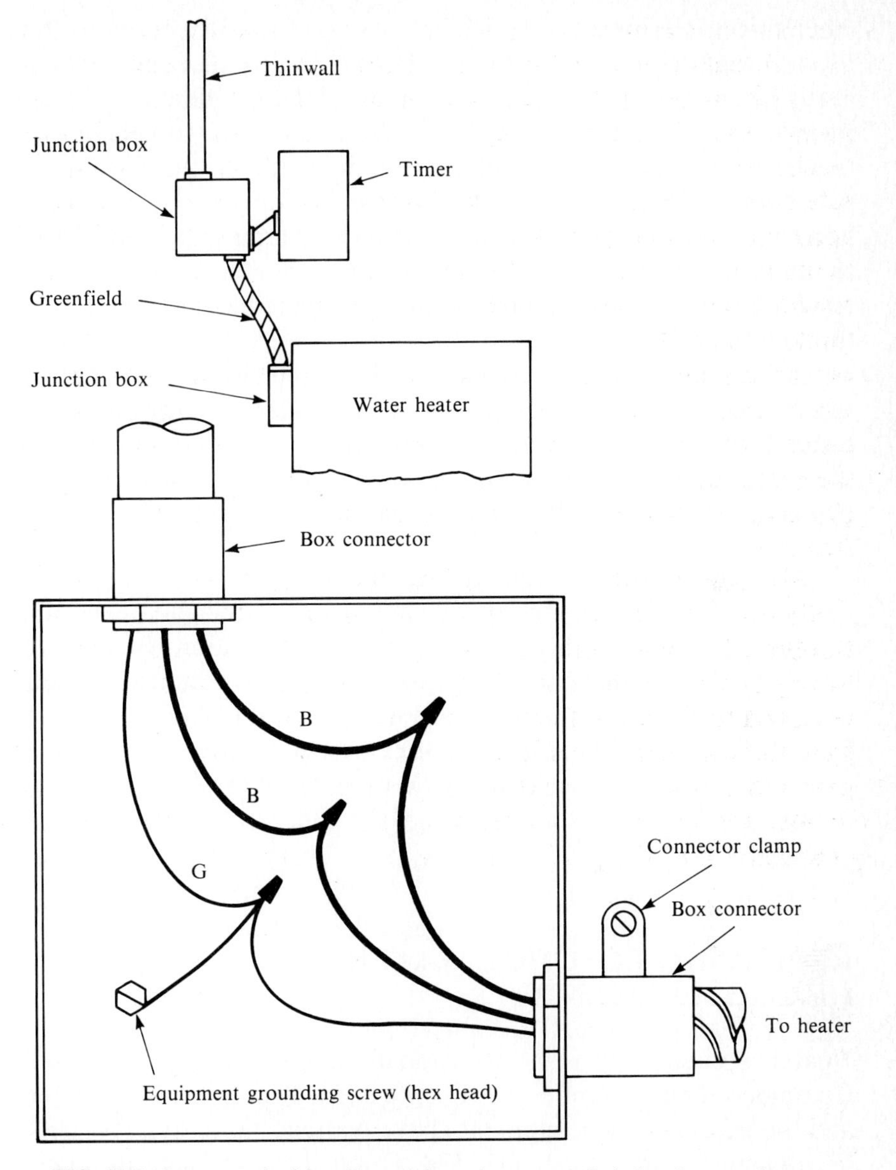

FIG. 10-7 Layout and wiring diagram for a timer-controlled electric water heater.

mechanism is removable from the case) and connect them to the *line* terminals. If the wires are too short to do this, disconnect them at the heater junction box, pull them out of the greenfield, and feed them through the offset nipple into the timer case. Cut them off, leaving *plenty* of length to make the connections to the line terminals. You will now need an additional length of No. 10 black insulated wire to run *two* leads from the two load terminals in the timer to the heater connections. Also you will need a length of bare No. 10 wire from the junction box on the wall to the heater metal frame (for grounding the heater).

If the timer needs a separate circuit for the clock, 15 A- 120 V, run that from the nearest available location. Be sure this power is on at all times, or else the clock will not keep correct time. Usually the heater needs only to be on for two 2-hour periods, morning and evening. The on periods can be lengthened, if necessary.

> **CAUTION:** Turn off the power before you start any work. Many water heaters in my area are on separate meters with the main breaker for the heater at the meter location. Be sure to check this. This means that turning off the main house breaker will *not* de-energize the water heater circuit. *Check this out carefully!* Additionally, if a separate circuit feeds the timer clock, be sure to put a warning notice inside the timer case that two circuits feed this device. Note the warning sign in Fig. 10-9.

REPLACING ELECTRIC WATER HEATER ELEMENTS

Heater elements burn out occasionally and need to be replaced. Common elements come in two styles, the 1-inch-thread screw-in and the bolt-on. These elements have wattage ratings of 750–6000 W, ranging in price from $11–$38. Often these are sold in home-improvement centers for less. The thermostat and heating element are shown in Fig. 10-10.

To replace a defective element, turn off the power. Then drain the tank. To facilitate drainage, open a hot-water faucet. The valve to turn off the water to the tank is on top of the tank in the cold-water line. For heaters in utility rooms, attach a hose to the drain cock and lead it outdoors. Again, check that the power is off.

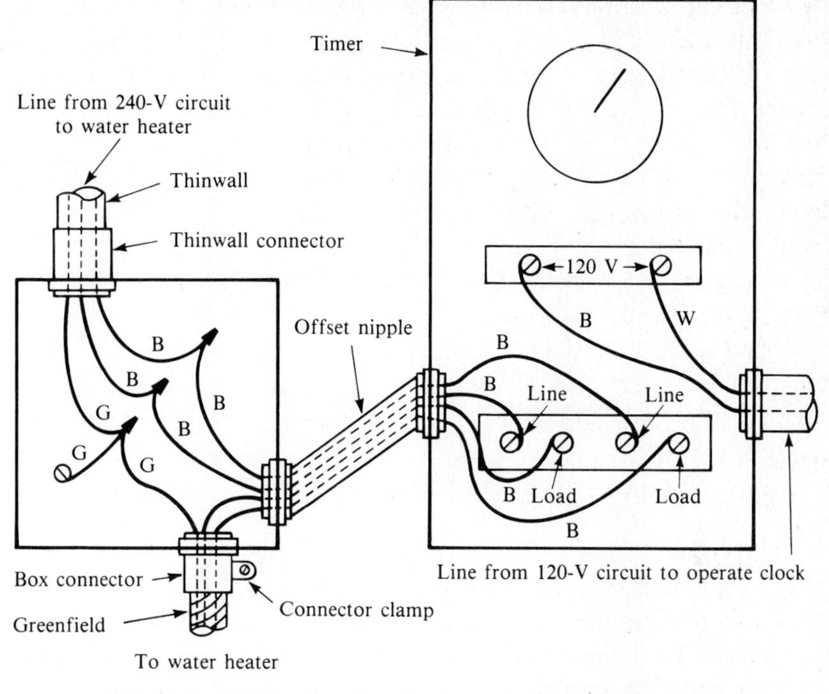

FIG. 10-8 Wiring connections for the water heater in Fig. 10-7.

Remove the cover plate to gain access to the element and its thermostat. You may have to remove the thermostat before or with the element. Disconnect the wires, and move them out of the way. Either unbolt or unscrew the element, and pull it out of the tank. The replacement element must be the same exact length as the old one (a longer element may contact the opposite side of the tank). Take the old element with you when you go to buy a replacement. Most replacement elements are universal and will fit many situations. Take the data from the heater nameplate with you so that you can get the same wattage.

Screw or bolt the replacement element in place.

> ***Note:*** The threaded element *must* have thread-sealing compound applied to its threads (such as teflon tape) before being inserted and tightened in place.

Tighten carefully. Fill the tank and check for leaks before reconnecting the wiring. At this time you can lower the thermostat setting to 120°F for energy saving in addition to the installation of a

FIG. 10-9 Water heater timer with two power sources.

timer. Another energy saver you might consider is an insulating water heater blanket.

REPLACING HEATING ELEMENTS AND DEVICES IN AN ELECTRIC RANGE

Electric range surface elements are very easy to replace; some just unplug, others must be disconnected from wires.

> ***Note:*** Before disconnecting any wires, unplug the range cord from its receptacle. Oven elements are similar, but are usually wired in and do not unplug. *Always* disconnect the cord for anything other than unplugging a surface element.

Take the part with you when you buy a replacement part, and have the nameplate date with you too. Be sure to have the brand of

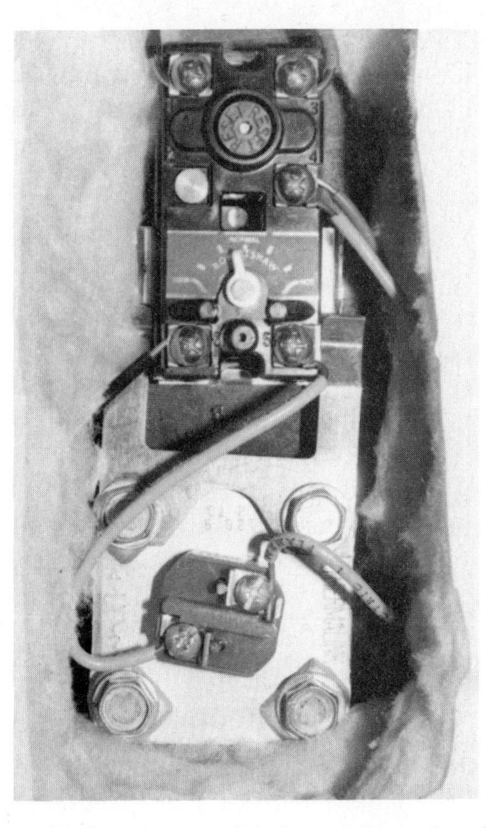

FIG. 10-10 The thermostat is at the top, and the heating element is at the bottom. Element is the bolt-on type.

range from the nameplate. The parts store needs as much information as possible.

Each part purchased must be the exact replacement for the old part. In rare cases the parts store will not be able to match the old part. Then you will have to go to the service center for that brand of range. Call first to find out whether the part is available. Service centers may charge more than parts stores, but this may be your only option.

Old ranges may have wiring which is in poor condition and should be replaced. This takes special wire, different from ordinary wire; this is available from the electrical supply house. To get the right type, tell them what the wire is to be used for. You may need crimp-on connectors and a crimping tool for the connections using the new wire. Make careful connections. Replace *each* wire singly, and remove only one wire at a time. If you are using only one color of wire, it would be wise to buy wrap-on labels (Brady or other brands) to attach to the new wires. These come in books having numbers from 1–25 or 1–50; each page has one number on up to 25 labels. There are also single squares with a number corresponding to the wrap-on on that page. This label is to be placed next to a terminal screw to identify that screw. Wrap-on numbers are shown in Fig. 10-11.

Many ranges have wrap-on designations which correspond to the wiring diagram attached to the rear panel back; others use colored wires. Be sure to make a proper list to transfer *colored* wires to *numbered* wires (new) which you are installing, such as no. 3 = blue; no. 4 = purple, etc. Numbered wires will remain the same. Most infinite switches (for the surface burners) and thermostats will be easy to find replacements for. Be sure to follow the diagram furnished with the new part, switch or thermostat, *exactly,* because either may be damaged by incorrect wiring.

REPLACING ELECTRIC PARTS IN A WASHER OR DRYER

Automatic washers are similar to electric ranges but are complicated by a timer which activates a multiple-contact switch to sequence the washer's operation. As described in Chapter 9 on motors, appliance motors are also subject to having dirt on the internal switch. Refer to Chapter 9 to clean these contacts. Under

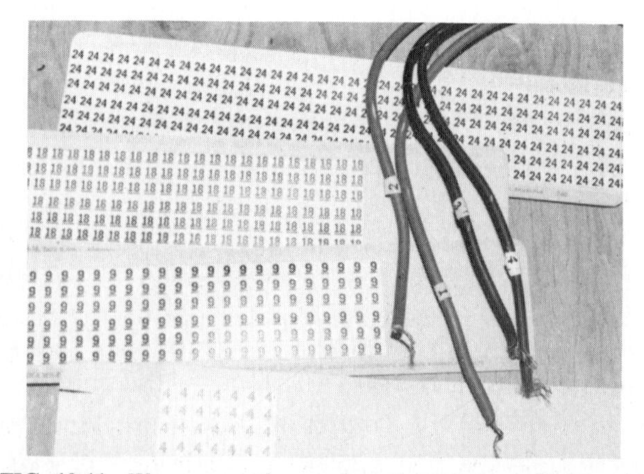

FIG. 10-11 Wrap-on number cards and numbers applied to wires.

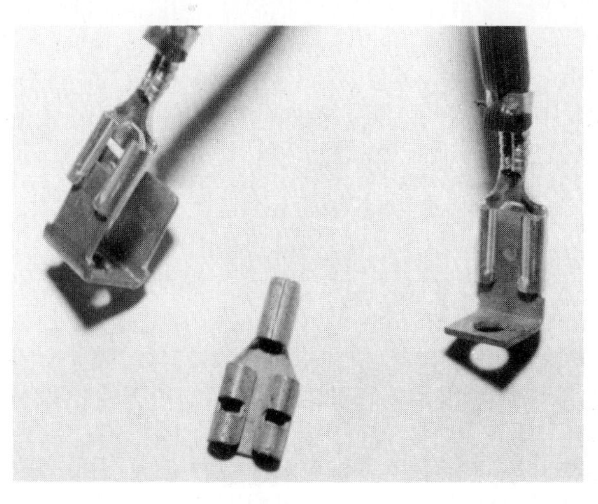

FIG. 10-12 Push-on crimp-style terminals. Both male and female types are shown.

these conditions, the motor will hum but will not start. This inability to start may burn the contacts on the multiple-contact switch. In any case you will need the wrap-on numbers because the cycle switch has many wires. Make wiring diagrams of everything, so that you will make no mistakes when replacing wires. As usual, disconnect the appliance from the receptacle *before* starting work.

Each appliance is designed differently, and removing panels and covers is a challenge in ingenuity. Take your time and do not force anything. When you remove controls, take care not to damage them. Repair manuals are available from most manufacturers; these give wiring diagrams and detailed instructions for making repairs. Terminals are shown in Fig. 10-12.

REPLACING FURNACE ELECTRIC COMPONENTS

Electric devices on a furnace or boiler are easily replaced if defective. Devices which may give trouble include the fan control or the limit control or a combination fan/limit control. Boilers have a circulator (pump) control and a limit control. There may be another control which keeps the circulator from running until the boiler water is heated.

The fan control can be tested for proper operation. The cycle of operation is as follows: The thermostat calls for heat, lighting the gas burner. The burner heats the furnace to about 130 to 140°F, depending on the fan control setting. When the furnace is heated to the setting, the fan will start. When the thermostat is satisfied, the burner will shut off; the fan will continue to run until the furnace cools down to about 90°F, depending again on the control setting. To check the control for proper operation, install a jumper wire across the terminals of the control (turn the furnace switch to off before doing this). Turn on the furnace switch; the fan should run even with the burner off and the furnace cold. If the fan hums, then the internal switch in the motor needs cleaning, or else the contacts are burned and the switch must be replaced. If the motor makes no sound, then there is a broken wire between the control and the motor. If the fan switch is defective, it must be replaced. With the power off, remove the switch. Remember to take it with you when you buy the new one. The length of the element insert is important because the element must not touch any part of the furnace metal. You will have to buy the fan switch from a heating wholesaler.

The furnace *limit* switch is designed to shut down the burner in a condition where the fan can't run because of a defective fan switch, a broken fan belt, a slipping belt (in this case, the fan would run too slowly and not remove the heat from the furnace), or squirrel-cage fan blades that are caked with dirt. In these cases the limit switch is doing its job in shutting down the burner to prevent overheating of the furnace.

The limit switch seldom goes bad, since it does not get the constant use that the fan switch does. Test the limit switch by turning off the furnace and removing one wire from the limit control. Turn the furnace back on and raise the thermostat setpoint (setting) to its highest setting. Now the furnace burner should not come on. Again turn off the furnace, replace the wire on the limit control, and turn the thermostat back to its normal setpoint (do not forget to return the thermostat to its normal setpoint, otherwise the house will overheat). This is correct wiring. Next, after turning off the furnace, remove one wire from the fan control. This will stop the fan from operating. Move the thermostat to its highest setting. Turn on the furnace; the furnace burner will come on and heat up the furnace. Since the fan cannot run, the burner will continue to heat the furnace until the limit control shuts down the burner at about 180–200°F furnace temperature. Turn off the furnace and replace the wire on the fan control.

CAUTION: Reset the thermostat. Turn the furnace back on. This test confirms that the limit control operates when needed. If the limit control will not shut off the burner, either the limit control is defective or the wiring is incorrect or defective.

Errors *do* occur; I once needed to do some work on air conditioning equipment. The compressor had a high- and low-limit control combined. When checking the compressor, I found it necessary to test the low-limit operation. As the unit ran, it continued running far past the control setting. I shut off the unit and checked the wiring. The control had *never* been wired into the system; the compressor could have been damaged by not being able to shut itself off as designed.

APPENDIX

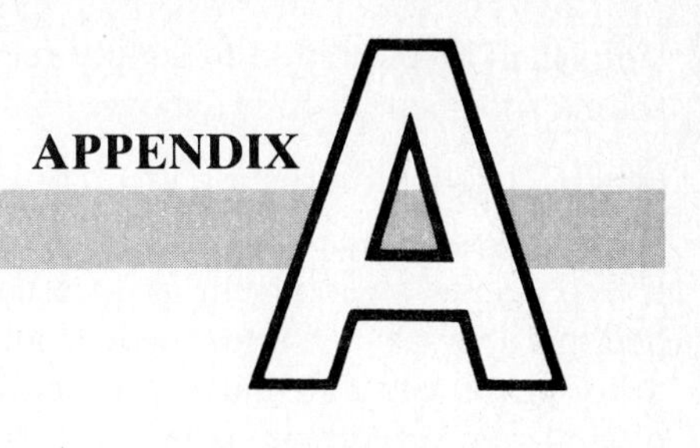

National Electrical Code® Revisions*

CHANGES INCORPORATED IN THE 1984 NEC†

Section 210-52 (d).

(d) Outdoor Outlets: For a one-family dwelling at least one receptacle outlet accessible at grade level shall be installed outdoors. For a two-family dwelling at least one receptacle outlet accessible at grade level shall be installed outdoors for each dwelling unit which is at grade level. (See Sec. 210-8 (a) (3).)

Section 220-1. (Paragraph 2)

Unless other voltages are specified, for purposes of computing branch circuit and feeder loads, nominal system voltages of 120, 120/240, 208Y/120, 240, 480Y/277, 480 and 600 V shall be used.

Section 220-3 (b) (2).

Countertop receptacle outlets installed in the kitchen shall be supplied by not less than two small appliance branch circuits, either or both of which shall also be permitted to supply receptacle outlets in kitchen and other rooms specified in (b) (1) above. Additional small appliance branch cir-

† Pertaining to the single dwelling.

cuits shall be permitted to supply receptacle outlets in kitchen and other rooms specified in (b) (1) above.

Section 240-4. Protection of Fixture Wires and Cords.

Flexible cord, including tinsel cord and extension cords, shall be protected against overcurrent in accordance with their ampacities as specified in Table 400-5. Fixture wire shall be protected against overcurrent in accordance with its ampacity as specified in Table 402-5. Supplementary overcurrent protection as in Sec. 240-10 shall be permitted to be an acceptable means for providing this protection.

Exception No. 1: When a flexible cord or a tinsel cord approved for and used with a specific listed appliance or portable lamps is connected to a branch circuit of Article 210 in accordance with the following.

20 A circuits	Tinsel cord or No. 18 cord and larger
30 A circuits	No. 16 cord and larger
40 A circuits	Cord of 20 A capacity and over
50 A circuits	Cord of 20 A capacity and over

Exception No. 2: When fixture wire is connected to 120-V or higher branch circuit of Article 210 in accordance with the following:

20 A circuits	No. 18 up to 50 ft (15.2 m) of run length
20 A circuits	No. 16 up to 100 ft (30.5 m) of run length
20 A circuits	No. 14 and larger
30 A circuits	No. 14 and larger
40 A circuits	No. 12 and larger
50 A circuits	No. 12 and larger

Exception No. 3: Flexible cord used in listed extension cord sets in lengths of 25 ft (7.62 m) or less and having No. 16 AWG conductors, or any length of larger conductors, shall be considered to be protected by 20-A branch-circuit overcurrent protection.

Section 250-71. Service Equipment

(a) Bonding of Service Equipment: The non-current-carrying metal parts of equipment indicated in (1), (2), and (3) below shall be effectively bonded together.

(3) Any metallic raceway or armor enclosing a grounding electrode conductor as permitted in Sec. 250-92a.

Section 250-74.

Exception No. 1: Where the box is surface mounted, direct metal-to-metal contact between the yoke and box shall be permitted to ground the receptacle to the box. This exception *shall not apply* to cover-mounted receptacles unless the box and cover combination are listed as providing satisfactory ground continuity between the box and the receptacle.

Section 250-114. Continuity and Attachment of Branch-Circuit Grounding Conductors in Box

Where more than one equipment grounding conductor enters a box all such conductors shall be splices or joined within the box or to the box with devices suitable for the use. Connections depending on solder shall not be used and the arrangement shall be such that disconnection or removal of a receptacle, fixture or other device fed from the box will not interfere with or interrupt the grounding continuity.

Section 300-5 (d). Underground Installations

(d) Protection from Damage: Conductors emerging from the ground shall be protected by enclosures or raceways extending from a minimum of 18 inches (457 mm) below grade to a point at least 8 ft (2.44 m) above finished grade.

Conductors entering a building shall be protected to the point of entrance.

Where the enclosure or raceway is subject to physical damage the conductors shall be installed in rigid metal conduit, intermediate metal conduit, Schedule 80 rigid nonmetallic conduit or equivalent.

Section 300-15 (b). Box Only

Exception No. 4: Where cables enter or exit from conduit or tubing which is used to provide cable support or protection against physical damage. A fitting shall be provided on the end(s) of the conduit or tubing, to protect the wires or cables from abrasion.

Section 336-2. Construction (Paragraph 2)

Conductors of Types NM and NMC shall be one of the types listed in Table 310-13 which is suitable for branch-circuit wiring or one which is identified for use in these cables. Conductors shall be rated at 90°C (194°F). The ampacity of Types NM and NMC cable shall be that of 60°C (140°F) conductors in Table 310-16.

Section 410-4 (d). Fixtures in Specific Locations

(d) Pendants: No parts of hanging fixtures or pendants shall be located within a zone measured 3 ft (914 mm) horizontally and 8 ft (2.44 m) vertically from the top of the bathtub rim. This zone is all encompassing and includes the zone directly over the tub.

Section 410-73 (e). General

(e) Thermal Protection: Where fluorescent fixtures are installed in-

doors, the ballasts shall have thermal protection integral within the ballast. Replacement ballasts for all fluorescent fixtures installed indoors shall also have thermal protection integral within the ballast. Exception to (e) above: Fluorescent fixtures with simple reactance ballasts.

Section 426-3. Application of Other Articles

All requirements of this code shall apply except as specifically amended in this article. Cord and plug connected fixed outdoor electric de-icing and snow-melting equipment intended for specific use and identified as suitable for this use shall be installed according to Article 422. Fixed outdoor electric de-icing and snow-melting equipment for use in hazardous (classified) locations shall comply with Articles 500 through 516.

Section 445-6. Protection of Live Parts

Live parts of generators operated at more than 50 V to ground shall not be exposed to accidental contact where accessible to unqualified persons.

Section 550-23 (d). Location

Mobile home service equipment shall be readily accessible and shall be located in sight from and not more than 30 ft (9.14 m) from the exterior wall of the mobile home it serves.

EXPLANATION OF THE CHANGES

Section 210-52 (d).

This section has now been changed to require one outdoor receptacle for *each* dwelling unit rather than only one for the entire building.

Section 220-1.

A second paragraph has been added to have the stated voltages conform to the actual voltages supplied by the utility.

Section 220-3 (b) (2).

Wording has been changed to minimize the overloading of circuits feeding receptacles in the counter area of kitchens.

Exception No. 2: These circuits may also supply an outdoor receptacle.

Section 240-4.

The protection requirements are revised because of the possibility of fire.

Section 250-71.

This clarifies that *any* type of raceway *must* be grounded. Where the grounding wire goes through an isolated piece of raceway (conduit), both ends of the raceway must be grounded to the grounding wire inside by using clamps and short sections of wire.

Section 250-74.

Exception No. 1: Receptacles attached to cover plates and designed for mounting on junction boxes are usually held to the cover plate *only* by the center mounting screw. It is therefore mandatory that a green grounding pigtail be installed from the receptacle green grounding screw to the box grounding screw.

Section 250-114.

Where more than one cable (Romex) enters a receptacle or switch box, all the bare ground wires must be twisted together and secured by a wire connector (wire nut), including a pigtail end to attach to the receptacle or other device to be mounted in the box. In this manner, removing the device will still preserve the continuity of the grounding system. All these connections *must* be made within the box.

Section 300-5.

This section requires that wires underground, such as from a dwelling to a garage, have conduit from the bottom of an 18-inch-deep ditch and then curving upright against the building wall, terminating with an angle fitting and going through wall and inside the building. When conduit goes up a pole, it must extend at least 8 ft above grade.

Section 300-15 (b).

Exception No. 4: Conduit used to protect cable (Romex) from damage must have bushings on each end to avoid abrasion. An example is a feed for a receptacle for a washing machine in the basement with Romex coming down the concrete or block wall, terminating in a wall box and receptacle. The bushing is used only at the top or ceiling end. The box protects the other end.

Section 336-2.

It is now required to have the insulation on cables such as Romex to be rated at 90°C (194°F). This is to prevent the deterioration of the insulation on wires covered with *building* insulation.

Section 410-4 (d).

Persons standing on the tub edge in a bathroom and touching a hanging light fixture may receive a fatal shock. This new rule prevents such a situation.

Section 410-73.

Replacement ballasts used in inside fluorescent fixtures must have integral thermal protection. This also applies to new fluorescent fixtures. Fire may ensue with ballasts that are not protected in this manner.

Section 426-3.

Ice- and snow-melting cables may now be plugged into standard outdoor receptacles (protected by GFCIs, as is also required), given approved cords and plugs are used for this purpose.

Section 445-6.

Auxiliary generating systems must have all live wires and other parts protected from accidental contact if they are 50 V and over.

Section 550-23 (d).

The service equipment disconnecting switch for a mobile home must be within sight of the near side of the home. This is to prevent someone turning this switch back on when another person is working on the home wiring.

APPENDIX B

Glossary

Accessible: Able to be removed or exposed without damaging the building structure or walls.

Ampacity: The current-carrying capacity of electric conductors, expressed in amperes.

Appliance: Utilization equipment (current-using) which may be either fixed (permanently wired to the system) or portable (plugged into a receptacle), for example, the toaster, electric range, water heater, clothes dryer.

Approved: Acceptable to the authority having jurisdiction.

Attachment Plug (Plug Cap, Cap): Device at the end of an appliance cord to connect an appliance to power.

Automatic: An electric unit that starts and stops on its own by means of some controlling device such as a thermostat.

Bonding: Joining two metal parts to form a solid connection electrically.

Bonding Jumper: A wire connection that joins two metal parts which must be joined at all times even though they may be separated at times. (For example, removal of the water meter breaks the continuity of the water pipe for grounding purposes.)

Bonding Jumper, Main: Jumper that connects the neutral bar in the service entrance panel to the metal of the panel to ground the panel.

Branch Circuit: The wires or cable from the fuse or breaker to the point of current usage (for example, a receptacle or ceiling fixture).

Branch Circuit, Individual: A circuit that has only one piece of utilization equipment connected to it (for example, the range or dryer).

Building: A separate structure standing alone or cut off from adjoining structures by solid walls or fire doors.

Cabinet: An enclosure that contains wires and devices such as breakers or fuses. (Can be surface- or flush-mounted.)

Circuit Breaker: An overcurrent device that automatically opens the circuit (or may be switched manually) to disconnect the circuit in case of overload or short circuit.

Close: To operate a switch to complete a circuit. Also called *make*.

Concealed: Not accessible without damage to the building or building surface; includes wires in conduit or tubing.

Conductor: Any wire or busbar that is capable of carrying electricity; may be bare or insulated.

Connector, Pressure (Solderless): A terminal end to be attached under a screw or pushed onto or into a mating terminal end. The wire is inserted in the hollow end and then is crimped to make a secure electric connection.

Continuous Load: Any load which is continuous for more than 3 hours.

Dead: Having no voltage; not attached to any source of current.

Dead Front: As applies to a cabinet or panel that has all live parts covered to prevent shock.

Device: Any unit carrying current but not utilizing current (for example, a switch, receptacle, or thermostat).

Die: A tool to cut threads on round stock such as a bolt.

Dwelling: House, home, living quarters for one or more persons.

Edison-Base Thread: The thread found on light bulbs and fuses.

Elevation: Height above a reference point such as a floor.

Feedback (Separate Feed): A source of current other than the obvious or usual source. The device may need two current sources turned off before it can be worked on safely.

Ferrule: A sleeve, usually brass, used to help seal waterproof fittings for flexible conduit.

Fill: The percentage of space in an electric box that wires may occupy without crowding. (An NEC requirement.)

Fish: To push or pull cable inside hollow walls or through attics while attempting to snag the end, retrieve it, and pull it through a wall opening. (Classed as *old work.*)

Fuse: The weak link in a circuit which fails, as it was designed to do, upon overload or short circuit, thereby protecting wires and equipment from getting too hot.

Fuse Block, Pull-Out Type: Used in 60-A and smaller service entrance panels. Has plastic block with fuse, fuse clips, and prongs. Prongs fit into a recessed receptacle which accepts the whole block, making a flush arrangement in the fuse box.

Fused Disconnect: A device having fuses in a cabinet; also an externally operated switch to disconnect the power from certain equipment.

Ground: A connection to earth, whether accidental or intentional; or an object that is connected to earth.

Ground*ed* Conductor: A conductor intentionally grounded.

Ground Fault Circuit Interrupter (GFCI): A device for protecting people against shock by detecting small current flow and breaking the circuit.

Ground*ing* Conductor: The wire from the equipment to the established ground (usually the water pipe).

Grounding Conductor, Equipment: The wire connecting the metal non-current-carrying parts (for example, the panels, cabinets, etc.).

Hickey: A tool for bending rigid conduit. Looks like a pipe T with one side cut away.

Hot: Energized wire, motor, or other piece of equipment. Another word is *live*.

ID: Inside diameter, as in a pipe or conduit.

In Sight From (Within Sight From): A disconnect located so that someone working on electric equipment can see whether the disconnect is still turned off.

Knife Blade: Refers to a large-amperage fuse having thin ends instead of round. Also refers to disconnect switches with similar blades.

Knockout: Round, partially punched-out openings in cabinets and junction boxes. These can be knocked out with a screwdriver.

Lighting Outlet: A connection to an outlet for holding a bulb or ceiling fixture.

Location, Damp: A location sheltered by overhanging awnings or roofs; barns and sometimes basements.

Location, Dry: A location not subject to dampness.

Location, Wet: Underground or in contact with water.

Locknut: A flat nut which screws onto a cable or conduit fitting to secure it in a knockout.

Meter Base: A casting or metal box having connections for the electric utility meter. The meter plugs in much as an attachment plug fits into a receptacle.

Mouse: A tool usually handmade by coiling up wire solder to form a small, heavy weight having a loop to tie on a line.

Nameplate Rating: A metal or plastic plate attached to motors, air conditioners, and household appliances which gives data regarding operating characteristics such as voltage and amperage.

New Work: Electrical work installed in unfinished buildings but having partition framing and enough other work installed that an electrician can complete the roughed-in wiring. This work will be concealed when the wall and ceiling finish is applied.

Normal: The condition of a switch or other piece of equipment when it has no external force acting on it. In other words, the condition of the device when you remove it from its shipping container.

Nut Driver: A screwdriver-type hand tool different in that it has a small socket instead of a screwdriver tip.

OD: Outside diameter, as in conduit or tubing.

Old Work: Installation of electric wiring in a finished building which makes it necessary to fish wire and cable inside finished wall by cutting and patching.

Open (*noun*): A break in the continuity of a circuit or wire, making it impossible for current to flow.

Open (*verb*): To operate a switch so that it breaks the circuit. The opposite of *close*.

Overload: Excess current flow greater than the device or equipment is designed for.

Panel: A cabinet with its component parts, such as breaker, fuse panel, or service equipment.

Parallel: Having two or more separate paths for electricity to travel.

Photocell: A device for sensing the absence of light, thereby closing internal contacts to turn lighting on during darkness.

Pry-Out Knockout: A knockout having a screwdriver slot to be used to pry or twist out the knockout without pounding it out.

Romex: A brand name for nonmetallic cable used in house wiring.

Sash Chain: Chain used to support double-hung sash instead of a sash cord. A short length is also used as a mouse (*see above*).

Series: Having only one path for electricity, as the older types of series Christmas lights. When one burns out, they all go out.

Series-Parallel: A combination of both wiring arrangements.

Service Drop: Overhead wires from the utility pole to the house.

Service Head: The special fitting at the top of the service mast or service entrance cable to keep out rain and point the wires down.

Service Lateral: Underground wires from the utility to the house.

Thermal Protector: An automatic protective device usually on or in a motor to protect it against burn-out.

Terms not Defined by the NEC

A.W.G.: American Wire Gauge; adopted to standardize wire sizes.

Bushing, Conduit: A screwed fitting for protecting wires leaving conduit; has rounded edges.

Bushing, Other: A fitting for reducing the size of a pipe thread to accept a smaller fitting; a fiber bushing to be inserted in the cut end of BX to protect the wire.

BX: Amored Bushed Cable; is similar to Romex in uses, but has a spirally wrapped metal casing for more protection. Very flexible.

Index